4 The *Wind Vision* Roadmap: A Pathway Forward

Summary

Chapter 4 and Appendix M provide a detailed roadmap of technical, economic, and institutional actions by the wind industry, the wind research community, and others to optimize wind's potential contribution to a cleaner, more reliable, low-carbon, domestic energy generation portfolio, utilizing U.S. manufacturing and a U.S. workforce. The roadmap is intended to be the beginning of an evolving, collaborative, and necessarily dynamic process. It thus suggests an approach of continual updates at least every two years, informed by its analysis activities. Roadmap actions are identified in nine topical areas, introduced below.

Wind Power Resources and Site Characterization

Significant reductions in the cost of wind power can be achieved through improved understanding of the complex physics governing wind flow into and through wind plants. Better insight into the flow physics has the potential to guide technology advancements that could increase wind plant energy capture, reduce annual operating costs, and improve project financing terms to more closely resemble traditional capital projects.

Wind Plant Technology Advancement

Technology advancements can provide increased energy capture, lower capital and operating costs, and improved reliability. Sustained focus on the wind power plant as an integrated system will provide the proper context for such advancements. Many technology improvements can be developed as straightforward extensions of previously successful technology trends, while others will be the result of new innovations.

Supply Chain, Manufacturing, and Logistics

Achieving the *Wind Vision Study Scenario* cost and deployment levels, while also maximizing economic value to the nation, will require a competitive domestic manufacturing industry and supply chain capable of driving innovation and commercialization of new technologies. Such technologies will enable cost-effective production, transportation, construction, and installation of next-generation wind plants on land and offshore.

Wind Power Performance, Reliability and Safety

Wind power is becoming a mainstream, widespread technology. With this progress, asset owner/operators, utilities, and the public expect wind plants to meet the same operational reliability as conventional generation sources. While enormous progress has been made in reliability and availability of systems, significant reductions in overall cost of energy can still be realized through better operations and maintenance (O&M) practices. This is especially true in the offshore environment, where maintenance costs are significantly higher due to more difficult access.

Wind Electricity Delivery and Integration

Successfully addressing power system integration issues, while still maintaining electric power system reliability, is critical to achieving high wind penetrations at reasonable costs. Key issues in this area relate to increased variability and uncertainty posed by wind power at various time scales. Methods for managing the power system with moderate-to-high wind penetrations have evolved, and will likely continue to evolve as more actual experience is gained with wind power plants. Utilization of wind forecasting in operational practice of power systems and advanced controls on wind turbines can help operators decide on appropriate reserve levels. In some cases, operators will be able to deploy wind turbine and wind plant response capabilities to help manage the power system. Experience and research demonstrate these approaches can be executed at reasonable cost if appropriate actions are taken.

Wind Siting and Permitting

As is true for any form of energy, wind power is associated with impacts to the natural surroundings. Wind is a comparatively clean source of energy with many positive attributes, such as no emissions, no air or water pollution, and no use of water in the generation of electricity. If improperly sited, however, wind power facilities may present socioeconomic, conflicting use, and environmental risks. Care needs to be taken in the siting of wind power facilities to ensure the potential for negative impacts from construction and operation is minimized to the greatest extent practicable.

Collaboration, Education and Outreach

Wind power development has experienced remarkable growth in terms of both deployment and technology innovation. The wind industry is seeing generational changes over the course of years, not decades, which can make it challenging for people not directly involved to stay abreast of this rapidly changing industry. Collaboration among domestic and international producers, researchers, and stakeholders during this time of rapid change facilitates learning about new approaches and technical advances that can lead to increased turbine performance, shorter deployment timelines, and lower overall costs.

Workforce Development

Realizing *Wind Vision Study Scenario* deployment levels and the associated benefits requires a robust and qualified workforce to support the industry throughout the product lifecycle. The industry needs a range of wind professionals, from specialized design engineers to installation and maintenance technicians, to enable the design, installation, operation, and maintenance of wind power systems. To support these needs, advanced planning and coordination are essential to educate a U.S. workforce from primary school through university degree programs.

Policy Analysis

Achieving wind power deployment to fulfill national energy, societal, and environmental goals—while minimizing the cost of meeting those goals—is likely to require practical and efficient policy mechanisms that support all three wind power markets: land-based, offshore, and distributed. Objective and comprehensive evaluation of different policy mechanisms is therefore needed, as are comparative assessments of the costs, benefits, and impacts of various energy technologies. Regular assessment of progress to enable ongoing prioritization of roadmap actions is also essential.

4.0 Introduction

Chapter 4 and Appendix M provide a detailed road-map of technical, economic, and institutional actions by the wind industry, the wind research community, and others to optimize wind's potential contribution to a cleaner, more reliable, low-carbon, domestic energy generation portfolio, utilizing U.S. manufacturing and a U.S. workforce. This roadmap was developed through a collaborative effort led by the U.S. Department of Energy (DOE), with contributions and rigorous peer review from industry, the electric power sector, non-governmental organizations, academia, national labs, and other governmental participants. High-level roadmap actions are presented and discussed in this chapter. Most of these actions are augmented by more detailed actions, which are described in Appendix M.[1]

The roadmap is not prescriptive. It does not detail how suggested actions are to be accomplished; it is left to the responsible organizations to determine the optimum timing and sequences of specific activities. While the *Wind Vision* report informs policy options, it is beyond the scope of the *Wind Vision* roadmap to suggest policy preferences and no attempt is made to do so.

The *Wind Vision Study Scenario* projects that wind technology costs continue to decline, that demand for wind power grows to support the *Study Scenario* penetration levels, and that wind power plants and the transmission assets needed to support them are actually built. In general, assumptions along these three lines are implicit in the *Study Scenario* modeling process. In aggregate, the roadmap actions are aimed at achieving the progress implied by these assumptions.

The Roadmap Approach

The *Wind Vision* roadmap outlines actions that can be taken by stakeholders under three distinct yet complementary themes designed to enable U.S. wind to compete for deployment in the U.S. power generation portfolio. The *Wind Vision* specifically does not make policy recommendations. By addressing market barriers, however, the *Wind Vision* roadmap actions can reduce the cost of complying with future proposed policy decisions and help improve market competitiveness of wind. The three key themes of the roadmap are:

1. **Reduce Wind Costs:** Chapter 3 demonstrates that the costs associated with the *Study Scenario* can be reduced across the range of sensitivities with wind cost reductions. Accordingly, reductions in levelized cost of electricity are a priority focus. This theme includes actions to reduce capital costs; reduce annual operating expenses; optimize annual energy production and reduce curtailment and system losses; reduce financing expenses; reduce grid integration and operating expenses; and reduce market barrier costs including regulatory and permitting, environmental, and radar mitigation costs.

2. **Expand Developable Areas:** Expansion of wind power into high-quality resource areas is also important for realizing the *Study Scenario* at cost levels described in Chapter 3 of the *Wind Vision* report. Key actions within this theme include actions to responsibly expand transmission and developable geographic regions and sites; improve the potential of low wind speed locales; improve the potential of ocean and Great Lakes offshore regions; and improve the potential in areas requiring careful consideration of wildlife, aviation, telecommunication, or other environmental issues. National parks, densely populated locations, and sensitive areas such as federally designated critical habitat are generally excluded from the roadmap actions, since they are likely not to be developed as wind sites.

---

1. The majority of the actions described in this chapter and Appendix M address utility-scale wind power, both land-based and offshore. DOE and the distributed wind community are assessing the prospects and development needs for distributed wind power. While several actions addressing distributed wind are included in the *Wind Vision* roadmap, the ongoing assessment is expected to generate a more complete set of distributed wind actions.

3. Increase Economic Value for the Nation: The *Study Scenario* projects substantial benefits for the nation, but additional steps are needed to ensure these benefits are realized and maximized. This theme includes actions to provide detailed and accurate data on costs and benefits for decision makers; actions to grow and maintain U.S. manufacturing throughout the supply chain; train and hire the U.S. workforce; provide diversity in the electricity generation portfolio; and provide a hedge against fossil fuel price increases. The overall aim is to ensure that wind power continues to provide enduring value for the nation.

The roadmap actions are intentionally limited to extensions and improvements of existing technologies and do not include transformational innovations. These innovations may occur, but, because of their novel nature, it is not possible to prescribe how to appropriately leverage them in advance in a roadmap process.

The roadmap is intended to be the beginning of an evolving, collaborative, and necessarily dynamic process. It thus suggests an approach of continual updates at least every two years, informed by its analysis activities. These periodic reviews will assess effects and redirect activities as necessary and appropriate through 2050 to optimize adaptation to changing technology, markets and political factors. High-level roadmap areas are summarized in Table 4-1 and the related high-level actions are summarized in Text Box 4-1. Appendix M provides a more granular description of the roadmap actions, including potential stakeholders and possible timelines for action.

Risk of Inaction

The analytical results of Chapter 3 reveal significant overall cumulative job, health, carbon, environmental, and other social benefits at deployment levels in the *Wind Vision Study Scenario*. Reduced economic activity and increased energy efficiency measures have slowed the growth of electricity demand and reduced the need for new generation of any kind. This decreased need for new generation, in combination with decreased natural gas costs and other factors, has reduced demand for new wind plants. These forces may cause the near-term U.S. market for wind equipment to fall below levels that will support a continued robust domestic manufacturing supply chain. If wind installation rates decline significantly, wind's ongoing contributions to U.S. economic development and U.S. manufacturing will likely be at risk. Wind operations will continue, but manufacturing will remain vibrant only as long as there are domestic markets to serve. If domestic markets for new installations deteriorate, manufacturing may move to other active regions of the world.

Table 4-1. *Wind Vision* Roadmap Strategic Approach Summary

Core Challenge	Wind has the potential to be a significant and enduring contributor to a cost-effective, reliable, low carbon, U.S. energy portfolio. Optimizing U.S. wind power's impact and value will require strategic planning and continued contributions across a wide range of participants.		
Key Themes	**Reduce Wind Costs** Collaboration to reduce wind costs through wind technology capital and operating cost reductions, increased energy capture, improved reliability, and development of planning and operating practices for cost-effective wind integration.	**Expand Developable Areas** Collaboration to increase market access to U.S. wind resources through improved power system flexibility and transmission expansion, technology development, streamlined siting and permitting processes, and environmental and competing use research and impact mitigation.	**Increase Economic Value for the Nation** Collaboration to support a strong and self-sustaining domestic wind industry through job growth, improved competitiveness, and articulation of wind's benefits to inform decision making.
Issues Addressed	Continuing declines in wind power costs and improved reliability are needed to improve market competition with other electricity sources.	Continued reduction of deployment barriers as well as enhanced mitigation strategies to responsibly improve market access to remote, low wind speed, offshore, and environmentally sensitive locations.	Capture the enduring value of wind power by analyzing job growth opportunities, evaluating existing and proposed policies, and disseminating credible information.
Wind Vision Study Scenario Linkages	Levelized cost of electricity reduction trajectory of 24% by 2020, 33% by 2030, and 37% by 2050 for land-based wind power technology and 22% by 2020, 43% by 2030, and 51% by 2050 for offshore wind power technology to substantially reduce or eliminate the near- and mid-term incremental costs of the *Study Scenario*.	Wind deployment sufficient to enable national wind electricity generation shares of 10% by 2020, 20% by 2030, and 35% by 2050.	A sustainable and competitive regional and local wind industry supporting substantial domestic employment. Public benefits from reduced emissions and consumer energy cost savings.
Roadmap Action Areas[a]	• Wind Power Resources and Site Characterization • Wind Plant Technology Advancement • Supply Chain, Manufacturing, and Logistics • Wind Power Performance, Reliability, and Safety • Wind Electricity Delivery and Integration • Wind Siting and Permitting • Collaboration, Education, and Outreach • Workforce Development • Policy Analysis	• Wind Power Resources and Site Characterization • Wind Plant Technology Advancement • Supply Chain, Manufacturing, and Logistics • Wind Electricity Delivery and Integration • Wind Siting and Permitting • Collaboration, Education, and Outreach • Policy Analysis	• Supply Chain, Manufacturing, and Logistics • Collaboration, Education, and Outreach • Workforce Development • Policy Analysis

a. Several action areas address more than one key theme.

1 Wind Power Resources and Site Characterization

Action 1.1 – Improve Wind Resource Characterization. Collect data and develop models to improve wind forecasting at multiple temporal scales—e.g., minutes, hours, days, months, years.

Action 1.2 – Understand Intra-Plant Flows. Collect data and improve models to understand intra-plant flow, including turbine-to-turbine interactions, micro-siting, and array effects.

Action 1.3 – Characterize Offshore Wind Resources. Collect and analyze data to characterize offshore wind resources and external design conditions for all coastal regions of the United States, and to validate forecasting and design tools and models at heights at which offshore turbines operate.

2 Wind Plant Technology Advancement

Action 2.1 – Develop Next-Generation Wind Plant Technology. Develop next-generation wind plant technology for rotors, controls, drivetrains, towers, and offshore foundations for continued improvements in wind plant performance and scale-up of turbine technology.

Action 2.2 – Improve Standards and Certification Processes. Update design standards and certification processes using validated simulation tools to enable more flexibility in application and reduce overall costs.

Action 2.3 – Improve and Validate Advanced Simulation and System Design Tools. Develop and validate a comprehensive suite of engineering, simulation, and physics-based tools that enable the design, analysis and certification of advanced wind plants. Improve simulation tool accuracy, flexibility, and ability to handle innovative new concepts.

Action 2.4 – Establish Test Facilities. Develop and sustain world-class testing facilities to support industry needs and continued innovation.

Action 2.5 – Develop Revolutionary Wind Power Systems. Invest research and development (R&D) into high-risk, potentially high-reward technology innovations.

3 Supply Chain, Manufacturing and Logistics

Action 3.1 – Increase Domestic Manufacturing Competitiveness. Increase domestic manufacturing competitiveness with investments in advanced manufacturing and research into innovative materials.

Action 3.2 – Develop Transportation, Construction, and Installation Solutions. Develop transportation, construction and installation solutions for deployment of next-generation, larger wind turbines.

Action 3.3 – Develop Offshore Wind Manufacturing and Supply Chain. Establish domestic offshore manufacturing, supply chain, and port infrastructure.

4 Wind Power Performance, Reliability, and Safety

Action 4.1 – Improve Reliability and Increase Service Life. Increase reliability by reducing unplanned maintenance through better design and testing of components, and through broader adoption of condition monitoring systems and maintenance.

Action 4.2 – Develop a World-Class Database on Wind Plant Operation under Normal Operating Conditions. Collect wind turbine performance and reliability data from wind plants to improve energy production and reliability under normal operating conditions.

Action 4.3 – Ensure Reliable Operation in Severe Operating Environments. Collect data, develop testing methods, and improve standards to ensure reliability under severe operating conditions including cold weather climates and areas prone to high force winds.

Action 4.4 – Develop and Document Best Practices in Wind O&M. Develop and promote best practices in operations and maintenance (O&M) strategies and procedures for safe, optimized operations at wind plants.

Action 4.5 – Develop Aftermarket Technology Upgrades and Best Practices for Repowering and Decommissioning. Develop aftermarket upgrades to existing wind plants and establish a body of knowledge and research on best practices for wind plant repowering and decommissioning.

High-Level Wind Vision Roadmap Actions *(continued)*

5 Wind Electricity Delivery and Integration

Action 5.1 – Encourage Sufficient Transmission. Collaborate with the electric power sector to encourage sufficient transmission to deliver potentially remote generation to electricity consumers and provide for economically efficient operation of the bulk power system over broad geographic and electrical regions.

Action 5.2 – Increase Flexible Resource Supply. Collaborate with the electric power sector to promote increased flexibility from all resources including conventional generation, demand response, wind and solar generation, and storage.

Action 5.3 – Encourage Cost-Effective Power System Operation with High Wind Penetration. Collaborate with the electric power sector to encourage operating practices and market structures that increase cost-effectiveness of power system operation with high levels of wind power.

Action 5.4 – Provide Advanced Controls for Grid Integration. Optimize wind power plant equipment and control strategies to facilitate integration into the electric power system, and provide balancing services such as regulation and voltage control.

Action 5.5 – Develop Optimized Offshore Wind Grid Architecture and Integration Strategies. Develop optimized subsea grid delivery systems and evaluate the integration of offshore wind under multiple arrangements to increase utility confidence in offshore wind.

Action 5.6 – Improve Distributed Wind Grid Integration. Improve grid integration of and increase utility confidence in distributed wind systems.

6 Wind Siting and Permitting

Action 6.1 – Develop Mitigation Options for Competing Human Use Concerns. Develop impact reduction and mitigation options for competing human use concerns such as radar, aviation, maritime shipping, and navigation.

Action 6.2 – Develop Strategies to Minimize and Mitigate Siting and Environmental Impacts. Develop and disseminate relevant information as well as minimization and mitigation strategies to reduce the environmental impacts of wind power plants, including impacts on wildlife.

Action 6.3 – Develop Information and Strategies to Mitigate the Local Impact of Wind Deployment and Operation. Continue to develop and disseminate accurate information to the public on local impacts of wind power deployment and operations.

Action 6.4 – Develop Clear and Consistent Regulatory Guidelines for Wind Development. Streamline regulatory guidelines for responsible project development on federal, state, and private lands, as well as in offshore areas.

Action 6.5 – Develop Wind Site Pre-Screening Tools. Develop commonly accepted standard siting and risk assessment tools allowing rapid pre-screening of potential development sites.

7 Collaboration, Education, and Outreach

Action 7.1 – Provide Information on Wind Power Impacts and Benefits. Increase public understanding of broader societal impacts of wind power, including economic impacts; reduced emissions of carbon dioxide, other greenhouse gases, and chemical and particulate pollutants; less water use; and greater energy diversity.

Action 7.2 – Foster International Exchange and Collaboration. Foster international exchange and collaboration on technology R&D, standards and certifications, and best practices in siting, operations, repowering, and decommissioning.

8 Workforce Development

Action 8.1 – Develop Comprehensive Training, Workforce, and Educational Programs. Develop comprehensive training, workforce, and education programs, with engagement from primary schools through university degree programs, to encourage and anticipate the technical and advanced-degree workforce needed by the industry.

9 Policy Analysis

Action 9.1 – Refine and Apply Energy Technology Cost and Benefit Evaluation Methods. Refine and apply methodologies to comprehensively evaluate and compare the costs, benefits, risks, uncertainties, and other impacts of energy technologies.

Action 9.2 – Refine and Apply Policy Analysis Methods. Refine and apply policy analysis methodologies to understand federal and state policy decisions affecting the electric sector portfolio.

Action 9.3 – Maintain the Roadmap as a Vibrant, Active Process for Achieving the Wind Vision Study Scenario. Track wind technology advancement and deployment progress, prioritize R&D activities, and regularly update the wind roadmap.

4.1 Wind Power Resources and Site Characterization

Significant reductions in the cost of wind power can be achieved through improved understanding of the complex physics governing wind flow into and through wind plants. Better insight into the flow physics has the potential to guide technology advancements that could increase wind plant energy capture [1], reduce annual operational costs, and improve project financing terms to more closely resemble traditional capital projects.

Realizing these opportunities will require diverse expertise and substantial resources, including high fidelity modeling and advanced computing. In order to validate new and existing high fidelity simulations, several experimental measurement campaigns across different scales will be required to gather the necessary data. These may include wind tunnel tests, scaled field tests, and large field measurement campaigns at operating plants. The data required include long-term atmospheric data sets, wind plant inflow, intra-wind plant flows (e.g., wakes), and rotor load measurements. Such measurement campaigns will be essential to addressing wind energy resource and site characterization issues and will yield improvements in models that bridge the applicable spatial-temporal scales.

Innovations in sensors such as Light Detection and Ranging, or LIDAR[2]; Sonic Detection and Ranging, or SODAR; wind profiling radars; and other new, high-fidelity instrumentation will also be needed to successfully collect data at the resolutions necessary to validate high-fidelity simulations. Significant effort will be required to store simulation results and data sets, enabling additional research and analysis in a user-friendly and publicly accessible database.

Resource characterization needs can be generally categorized in terms of increasing temporal and spatial scales:

- **Turbine dynamics**—representing phenomena such as turbulence and shear at the scale important to individual turbine inflow;
- **Micro-siting and array effects**—characterizing complex localized flows including terrain and turbine wake effects to optimize siting of turbines within projects and accurately predict power output;
- **Mesoscale processes**—representing wind fields across regional or mesoscale areas in the actual and forecasting timeframes important for operation of the electric power grid; and
- **Climate effects**—accounting for the effects of climate change and climate variability to protect long-term investments in wind power and other infrastructure.

More and better observations, improved modeling at all four spatial and temporal scales, and an integrated bridging of the four spatial-temporal scales are needed. *Research Needs for Wind Resource Characterization* [2] describes this topic in more detail.

The following actions focus on three key objectives:

- **Improving fundamental wind resource characterization to reduce the error and uncertainty of wind forecasts;**
- **Understanding intra-plant flows to optimize wind plant output; and**
- **Using data and analysis to devise offshore-specific wind resource characterization to better understand marine design and operating conditions.**

Detailed activities and suggested timelines for action are identified for each of these key areas in Appendix M.1.

---

2. Remote sensing technology that measures distance by illuminating a target with a laser and analyzing the reflected light.

Action 1.1: Improve Wind Resource Characterization

Improved characterization and understanding of wind resources are essential to increasing wind plant revenue and operating cost performance, thereby reducing risks to developers. This will contribute to reducing the cost of wind power and improve cost competition in the electricity sector.

ACTION 1.1: Improve Wind Resource Characterization	
Collect data and develop models to improve wind forecasting at multiple temporal scales—e.g., minutes, hours, days, months, years.	
DELIVERABLE	**IMPACT**
Data, validated models, and measurement techniques that improve ability to predict wind plant power output over several spatial and temporal scales.	Increased wind plant performance resulting in increased revenue, improved reliability, and decreased operating costs.
Key Themes: Reduce Wind Costs; Expand Developable Areas **Markets Addressed:** Land, Offshore	

Reducing the error and uncertainty of wind resource forecasts and wind power generation facilitates integration of wind into the electric grid. Stakeholder action is needed to develop, validate, and apply models and measurement techniques that accurately characterize and forecast the wind in various time frames—e.g., minutes, hours, days, months, years. Forecasts on the hourly scale support dispatch decisions, multi-hour forecasts warn of ramp events (rapid changes in power output), and day-ahead forecasts inform unit-commitment decisions. Two primary aspects of Numerical Weather Prediction are the data assimilation scheme and the model physics, both of which can be improved through stakeholder action to support wind integration.

Action 1.2: Understand Intra-Plant Flows

An improved understanding of the aerodynamic environment within a wind plant, including turbine-to-turbine interaction, is needed to optimize wind plant power production. Incorrect simulations of intra-plant flows can indirectly result in wind plant energy losses from wakes, complex terrain, and turbulence, as well as unknown turbine loading conditions that can result in over-designed turbine components.

High performance computational capability has recently become available to explore these issues [3], providing major opportunities to gain insight through advanced, high-fidelity modeling. For example, the High Performance Computing Data Center at the DOE's National Renewable Energy Laboratory provides high-speed, large-scale computer processing to advance research on renewable energy and energy efficiency technologies. Through computer modeling and simulation, researchers can explore processes and technologies that cannot be directly observed in a laboratory or that are too expensive or too time-consuming to be conducted otherwise [4].

ACTION 1.2: Understand Intra-Plant Flows	
Collect data and improve models to understand intra-plant flow, including turbine-to-turbine interactions, micro-siting, and array effects.	
DELIVERABLE	**IMPACT**
Data, validated models and measurement techniques to minimize turbulence induced by adjacent turbines through optimized siting.	Increased wind plant energy production and reduced turbine maintenance requirements.
Key Themes: Reduce Wind Costs **Markets Addressed:** Land, Offshore	

Action 1.3: Characterize Offshore Wind Resources

Meteorological and oceanographic (metocean) data are integral to defining the design and operating conditions over the lifetimes of offshore wind projects in regions where they may be sited. Construction and maintenance planning is dependent on known sea states (wave height, period, and power spectrum) and on the availability of accurate forecasts. In the United States, observational metocean data are sparsely collected and relies heavily on surface weather buoys that cannot probe hub height wind conditions. Significant portions of the oceans and Great Lakes lack year-round observational data and rely on models to estimate metocean conditions. These gaps cause a high level of uncertainty associated with the offshore resource and design environment; this in turn imposes additional cost and risk to offshore wind projects. Characterizing the metocean conditions through a series of measurement and modeling approaches targeted specifically for offshore wind power applications and the broad stakeholder community will help address these challenges.

ACTION 1.3: Characterize Offshore Wind Resources	
Collect and analyze data to characterize offshore wind resources and external design conditions for all coastal regions of the United States, and to validate forecasting and design tools and models at heights at which offshore turbines operate.	
DELIVERABLE	**IMPACT**
Resource maps, forecasting tools, weather models, measurement stations, and technical reports documenting physical design basis.	Improved offshore R&D strategy and accelerated offshore wind deployment.

Key Themes: Reduce Wind Costs; Expand Developable Areas
Markets Addressed: Offshore

4.2 Wind Plant Technology Advancement

Technology advancements can provide increased energy capture, lower capital and operational costs, and improved reliability. Sustained focus on the wind power plant as an integrated system will provide the proper context for these advances in technology. Many of these advances can be developed as straightforward extensions of previously successful technology trends, while others will be the result of new innovations.

The time and cost required to develop and certify new technology are substantial. To be successful, a consistent effort with many contributors is required. This reinforces the value of domestic and international partnerships that can bring together the resources necessary to fully realize opportunities for advanced technology.

Five key actions will support technology advancement:

- **Developing advanced wind plant sub-systems such as larger rotors;**

- **Updating design and certification standards to improve the certification process;**

- **Developing and validating comprehensive simulation tools to guide wind plant technology development;**

- **Developing and sustaining publicly available test facilities to verify the performance and reliability of new technology; and**

- **Devising a structured process to systematically identify and develop revolutionary concepts and invest R&D into potentially high-reward innovation.**

Detailed activities and suggested timelines for action are identified for each of these key areas in Appendix M.2.

Action 2.1: Develop Next-Generation Wind Plant Technology

The substantial advances in wind plant technology since 2008 illustrate the importance of technological innovation. Continued advances in technology will provide lower costs for wind power, increased deployment opportunities at lower wind speed sites, new offshore technology for both shallow and deep water, and improved reliability.

Innovations are needed that facilitate continued growth in the size and capacity of wind turbine systems. Opportunities for advancement exist in rotors, control systems, drivetrains, towers, and offshore foundations.

ACTION 2.1: Develop Next-Generation Wind Plant Technology	
Develop next-generation wind plant technology for rotors, controls, drivetrains, towers, and offshore foundations for continued improvements in wind plant performance and scale-up of turbine technology.	
DELIVERABLE	**IMPACT**
Wind power systems with lower cost of energy.	Reduced energy costs for U.S. industry and consumers. Increased wind deployment nationwide.

Key Themes: Reduce Wind Costs; Expand Developable Areas
Markets Addressed: Land, Offshore

Much larger rotors per installed megawatt (MW) are needed to continue expanding the range of sites across the United States that can produce wind power at a competitive price. Rotor blades that can be delivered to a wind plant in two or more pieces and assembled on-site will enable the continued growth of rotor diameters. Additional opportunities for innovation in the design of blades include aeroelastic design techniques that shed loads; advanced low-cost, high-strength materials; and active or passive aerodynamic and noise control devices.

Sophisticated turbine control systems will continue to contribute to increases in energy capture and the reduction of structural loads. Techniques that measure the wind upstream of individual turbines and wind plants, such as LIDAR, provide more accurate information about the flow field and allow control systems to take action before changes in the wind reach the turbines. Important opportunities are available in aerodynamic control to reduce structural loads using independent blade pitch control, as well as aerodynamic devices along the span of the blade. Wind plant control systems are evolving to operate the plant in an integrated manner, ensuring maximum energy capture, highest overall plant reliability, and active control of the plant's electrical output to provide ancillary services and support grid stability and reliability. In many cases, these improvements can be added with little or no increase in cost.

Continued development is needed to reduce cost and improve reliability and efficiency of the drivetrains and power conversion systems that turn the rotor's rotational power into electrical power. Conventional multi-stage geared approaches, medium-speed systems, and direct-drive architectures each have advantages, and technological development of all three configurations should be continued. High-flux permanent magnets can improve the efficiency of these configurations, and efforts to develop alternatives to the existing rare-earth technologies should be undertaken. New materials for power conversion electronics, such as silicon carbide, can increase efficiency and eliminate the need for complex liquid cooling systems.

Taller towers are the necessary complement to larger rotors. Taller towers also provide access to the stronger winds that exist at higher elevations and are a key enabler for cost-effective development of lower wind speed sites. Logistics constraints, however, limit the maximum diameter of tower sections that can be transported over land, and this causes the cost of tall towers to increase disproportionately. Innovations that permit increased on-site assembly of towers are needed.

The fabrication and installation costs of offshore foundations and support structures are higher than comparable costs for land-based wind. Offshore costs can be lowered considerably by reducing construction time and dependency on costly heavy-lift vessels, as well as through technology innovations, mass production, and standardization of the support structure. This opportunity will guide the development of advanced offshore foundations and substructures.

More than 60% of the gross U.S. resource potential for offshore wind is over water deeper than 60 meters [5]. At these depths, the cost of fixed-bottom substructures increases rapidly in comparison to shallower waters. Floating offshore wind turbine platforms may be more cost effective in these deep waters. New floating platform technologies should be developed with equivalent or lower costs than those of existing fixed-bottom systems. New technologies to mitigate high dynamic loading on the tower and support structure (imposed by flowing winter ice sheets) can enable development of up to 500 gigawatts (GW) of gross offshore wind resource potential in the Great Lakes. Hurricanes frequently affect the U.S. coastline from Cape Cod, Massachusetts, to Galveston, Texas, as well as Hawaii—affecting more than 1,000 GW of gross offshore wind resource area. New design approaches and operating strategies should be developed that can protect offshore wind turbines and foundations against these extreme events.

Action 2.2: Improve Standards and Certification Processes

The development in the 1990s of widely accepted international standards for the certification of wind power systems led to increased reliability, lower costs, and improved investor confidence. This was a key advancement that enabled the large-scale wind deployment that has occurred since the early 2000s.

ACTION 2.2: Improve Standards and Certification Processes	
Update design standards and certification processes using validated simulation tools to enable more flexibility in application and reduce overall costs.	
DELIVERABLE	**IMPACT**
Certification processes that provide the required level of reliability while remaining flexible and inexpensive.	Lower overall costs, increased reliability, and reduced barriers to deployment.

Key Themes: Reduce Wind Costs
Markets Addressed: Land, Offshore, Distributed

These standards, however, were developed in an era when simulation tools for wind power systems had limited capabilities. Turbine technology has evolved substantially since then, from relatively simple constant-speed, stall-regulated turbines that produced a few hundred kilowatts to modern, vastly larger and more sophisticated multi-MW turbines that use variable speed and full-span pitch control to limit structural loads and improve performance. This history has led to conservatism in the present generation of standards that may increase costs unnecessarily.

Making reliability a foundation of the next generation of standards should provide designers and manufacturers the flexibility to alter systems for optimization in specific sites, without excessive recertification costs or delays. Updated certification standards can be developed using a comprehensive process that measures structural loads and validates the accuracy of the industry's simulation tools for the full range of operational conditions experienced over the lifecycle of wind systems. These simulation tools will need to properly account for the operational environment in the interior of a wind plant using the complex flow data described in Section 4.1, *Wind Power Resources and Site Characterization*.

As of 2013, industry standards for offshore wind did not specifically address the design of structures in regions prone to tropical cyclones (hurricanes). This is a potential impediment to the development of more than 1,000 GW of offshore resources. New standards and engineering approaches are needed to design reliable offshore wind plants in light of risk of exposure to hurricanes. Combined wind, wave, and current data should be gathered during the passage of a hurricane to better define the operational environment. Detailed simulations of the structural response of the wind turbine and foundation during the passage of hurricanes should be completed to inform comprehensive design standards, and aerodynamic and structural data should be gathered on an operational turbine in a hurricane to validate these simulation results. Research is needed on the effect of hurricane conditions on wind plant capital and operational costs in order to optimize structural reliability provided by the standards and minimize lifecycle costs.

Developing and achieving consensus on revised international standards for the certification of wind power systems takes many years. Development of the next generation of standards will require collaboration among the wind industry, research laboratories, and national authorities around the world. Coordination with national, state, and local permitting processes is required to harmonize the new standards with the many authorities having jurisdiction over the permitting process.

Action 2.3: Improve and Validate Advanced Simulation and System Design Tools

Wind power system simulation tools, which are vital to modern engineering design and analysis, continue to see dramatic improvements in capability and accuracy. These improvements have reduced product development time and cost by largely eliminating the need for extensive redesign after initial testing. The need for these computer-aided physics and engineering tools to help reduce cost, increase reliability, and extend system lifetime will continue as wind deployment accelerates.

ACTION 2.3: Improve and Validate Advanced Simulation and System Design Tools	
Develop and validate a comprehensive suite of engineering, simulation, and physics-based tools that enable the design, analysis and certification of advanced wind plants. Improve simulation tool accuracy, flexibility, and ability to handle innovative new concepts.	
DELIVERABLE	**IMPACT**
Reliably accurate predictions of all characteristics of existing and novel wind turbine and wind plant configurations.	Improved technical and economic performance, increased reliability, and reduced product development cycle time.

Key Themes: Reduce Wind Costs
Markets Addressed: Land, Offshore, Distributed

A first step should be a comprehensive validation campaign to define the accuracy, strengths, and weaknesses of today's simulation tools for a wide range of modern wind power systems. This effort is directly supportive of the work described in Action 2.2 to develop next-generation certification standards, and will simultaneously support the identification of key opportunities and needs for improvements in the suite of simulation tools. Academia, research laboratories, and the wind industry can then collaborate to develop the improvements identified in the validation campaign. Rigorous assessment of simulation capabilities should be repeated periodically as the size and configuration of next-generation wind power systems evolve.

The focus of simulation tool development has been moving away from individual turbines and toward the challenge of simulating a complete wind plant (i.e., Simulator for Wind Farm Applications, or SOWFA; and Wind-Plant Integrated System Design & Engineering Model, or WISDEM).[3] This trend should be continued and accelerated. Improved simulation capabilities for the wind plant as a system enable further reductions in cost as well as improvements in performance. The ability to address the entire wind plant has been enabled by continued advances in capability and reductions in cost for high-performance computing systems.

Simulation tools under development can facilitate design of wind plants with significant gains in energy production and reduced structural loads. Flexible simulation tools should be created to support the design and development of innovative configurations for wind power systems. The ability to reliably simulate the operation of new configurations is a foundational capability for a successful innovation process.

Action 2.4: Establish Test Facilities

The performance and reliability of wind power systems need to be verified prior to deployment. This verification process is a key element of the overall risk management strategy, and is a requirement for certification. Test facilities also play an essential role in the development of new technologies. Current test facilities provide substantial capability, but increased capabilities are needed to support the continued growth of wind power.

---

3. Learn more about SOWFA at *https://nwtc.nrel.gov/SOWFA*, and more about WISDEM at *http://www.nrel.gov/wind/systems_engineering/models_tools.html.*

ACTION 2.4: Establish Test Facilities

Develop and sustain world-class testing facilities to support industry needs and continued innovation.

DELIVERABLE	IMPACT
Cost-effective, publicly available test facilities for all critical wind plant subsystems.	Lower cost of energy from increased reliability, reduced product development time, and support of innovative technology development.

Key Themes: Reduce Wind Costs
Markets Addressed: Land, Offshore

Wind power systems have grown to a scale at which test facilities become costly to develop and maintain. Shared facilities provide a cost-effective capability for the entire industry. Recently established facilities for the testing of blades (the Massachusetts Clean Energy Center's Wind Technology Testing Center) and drivetrains (dynamometers at the National Renewable Energy Laboratory and Clemson University) provide excellent examples of this approach.

As turbines have grown in size and capacity, it has become challenging to conduct field testing at appropriate sites. Industry manufacturers often use privately owned facilities for testing. In addition, the National Wind Technology Center near Boulder, Colorado, provides field testing at a site with strong and turbulent winds that are well characterized. The National Wind Technology Center site also has a Controllable Grid Interface that allows for the testing of wind turbines when anomalous grid conditions are present. Expansion of this field testing site would be necessary to meet the needs of the coming generation of much larger wind turbines. The field testing facility at Texas Tech University near Lubbock, Texas, provides a complementary field testing capability with lower turbulence and without the complex terrain of the National Wind Technology Center, and further expansion of this site would support larger turbines.

The Scaled Wind Farm Technology facility at the Texas Tech University site, known as SWiFT, provides an important capability to test turbine-to-turbine aerodynamic interactions at a small scale. Establishment of a similar facility to support testing at full scale would support continued wind growth. Installing research-grade instrumentation at an existing operational wind plant could provide an efficient approach for establishing a full-scale facility for measuring aerodynamic interactions. There is also a need for test facilities to validate innovative wind turbine and wind plant technologies at ultra-large scales and under unique offshore environmental conditions.

Other wind system test facilities exist in addition to those above. No publicly available facilities exist in the United States, however, to support testing of the many subsystems in a wind turbine and wind plant. The complex interactions between subsystems are frequently the root cause of reliability issues. A dedicated test facility could provide the capabilities to examine these interactions in a realistic and controlled environment.

Action 2.5: Develop Revolutionary Wind Power Systems

In the early days of the wind industry, a wide variety of configurations was developed and deployed. These configurations included both upwind and downwind rotors, various numbers of rotor blades, horizontal and vertical axis machines, and a variety of power conversion subsystems. Turbines with a horizontal axis rotor and three blades operating upwind of a tubular tower, with full-span blade pitch control and variable rotor speed, became the dominant configuration because of the excellent performance and reliability provided by this arrangement. Several decades of advancement in this configuration has given it a strong position in the marketplace.

Alternative configurations have been proposed that could provide advantages over the existing technology. Examples of these new configurations include floating vertical axis turbines, shrouded horizontal axis turbines, turbines with rotors that operate downwind of the tower, and airborne wind power systems. The cost and time to develop new wind turbine configurations at MW-scale, however, make it difficult for such innovations to enter the marketplace.

Public-private partnerships could be created to facilitate the multi-year, high-risk development process needed for these new technologies. A structured innovation process that identifies and provides solutions for the fundamental wind power conversion processes has the potential to provide a robust framework for this effort. Promising technologies should be demonstrated at increasing scale in a series of laboratory and field tests. The development of flexible simulation tools discussed previously is a critical enabler for this innovation. To ensure long-term success, support for this effort would need to transition from public to private sources as commercial prospects grow.

4.3 Supply Chain, Manufacturing, and Logistics

Achieving the *Wind Vision Study Scenario* for cost and deployment while also maximizing the economic value to the nation will require a competitive domestic manufacturing industry and supply chain capable of driving innovation and commercialization of new technologies. Such technologies will enable cost-effective production, transportation, construction, and installation of next-generation wind plants on land and offshore.

As described in Chapter 2, the U.S wind industry has enjoyed robust growth. An average of 7 GW per year was installed during the period from 2007 to 2013, hitting a peak of 13 GW installed in 2012, while the share of domestically produced components simultaneously increased. These manufacturing and installation trends demonstrate sufficient capability in the global wind industry to meet the levels of deployment presented in the *Wind Vision*. What is less certain is where next-generation wind power technology will be developed, where it will be manufactured, and whether the necessary infrastructure and technology will be available to transport and construct it both on land and offshore. Capturing the economic value to the nation will require a collective set of actions to be taken by a variety of stakeholders including industry, government, and academic and other research institutions to enhance and sustain a globally competitive domestic supply chain.

The following are key actions that build on the past accomplishments of the U.S. manufacturing, transportation, and construction industries:

- **Invest in advanced manufacturing and research in order to increase domestic manufacturing competitiveness;**

- **Develop transportation, construction, and installation solutions that support next-generation wind technologies; and**

- **Establish a domestic offshore manufacturing and supply chain.**

Detailed activities and suggested timelines for action are identified for each of these key areas in Appendix M.3.

Action 3.1: Increase Domestic Manufacturing Competitiveness

Given the continued increase in the size of wind components and the associated transportation limitations, the primary driver for a sustainable domestic supply chain is sufficient and consistent domestic demand for wind power—a key characteristic of the *Wind Vision Study Scenario*. Given this need for strong domestic demand, the question then becomes, "How can the U.S. supply chain capture the most economic value?" In industries with a global supply chain, such as wind, the value capture that a country realizes depends on the level of domestic content as shown in Chapter 2, section 2.4 of the *Wind Vision* report (*Economic and Social Impacts of Wind for the Nation*). This value also depends on the headquarters and R&D locations of the turbine and component manufacturers, who bring added profit margin on the product, provide additional jobs, and serve as a source of innovation to build and maintain global competitiveness [6].

ACTION 3.1: Increase Domestic Manufacturing Competitiveness	
Increase domestic manufacturing competitiveness with investments in advanced manufacturing and research into innovative materials.	
DELIVERABLE	**IMPACT**
New information, analysis tools, and technology to develop cost-competitive, sustainable, domestic wind power supply chain.	Reduced capital cost components, increased domestic manufacturing jobs and capacity, increased domestic technological innovation, and economic value capture.

Key Themes: Reduce Wind Costs; Increase Economic Value for the Nation
Markets Addressed: Land, Offshore, Distributed

The large size of major wind components makes them subject to transportation constraints and therefore suitable for domestic manufacture. Improving the competitiveness of the U.S. supply chain requires thorough analysis to understand the cost structure of U.S. and foreign suppliers, as well as assessment of the global trade and manufacturing policies that drive a majority of competitive differences. The data and tools developed can inform new policies that support U.S. manufacturers, and help industry prioritize key investments in manufacturing R&D and capital expenditures that can improve domestic manufacturing competitiveness [7, 8].

The results of the competitive analysis discussed here can be used to improve the cost structure of U.S. manufacturers and foster innovative manufacturing technology. Industry-led consortia, such as the Institutes for Manufacturing Innovation (established through the White House's National Network for Manufacturing Innovation), could serve as valuable forums to exchange knowledge, facilitate innovation, and develop technologies across industries and institutions that do not otherwise collaborate.

To advance domestic manufacturing competitiveness, the most promising new manufacturing technologies need to be scaled up and commercialized. Deploying new manufacturing technologies requires access to capital, which can be a significant barrier to U.S. manufacturers, especially the small and medium-sized businesses that make up a significant portion of the domestic wind supply chain. Analysis tools are needed to inform effective financial policies that enable domestic manufacturers to match and exceed the capabilities and capacity of foreign competition and manufacture the quantity, quality, and physical scale of next-generation wind plant technology [9].

Much of the cost of manufacturing is embedded in the raw materials and sub-assemblies that serve as inputs to the top tier manufacturers. Opportunities exist in steel mills, foundries, and fiber-reinforced polymer composite material suppliers to produce new standardized material forms that can reduce costly, high-labor content processes like welding or hand layup of composites that put domestic manufacturers at an inherent disadvantage due to the higher cost of labor in the United States. Coordinating with synergistic industries like aerospace, automotive, and offshore oil and gas would incent material suppliers with a diverse and sufficient market to retool or expand capacity. Best practices for wind turbine and wind plant decommissioning should also be developed in cooperation with the international wind industry to address the large amount of materials including steel and composites that will need to be recycled or dealt with in other ways as more plants reach the end of operation.

With respect to distributed wind, U.S. manufacturers dominate the domestic market for small wind turbines (i.e., through 100 kilowatts in size) and regularly export a noteworthy number of turbines [10]. In order for these manufacturers to remain competitive in the global market, continued investment in U.S. distributed wind turbine manufacturing technology and supply chain R&D is needed.

Successful implementation of these actions is expected to lead to an increasingly competitive domestic supply chain capable of achieving the deployment and cost goals of the *Wind Vision Study Scenario* and maximizing economic value for the nation. A competitive domestic manufacturing industry will also help develop and commercialize the many technologies described in Section 4.2, *Wind Plant Technology Advancement*. In addition to serving the domestic market, a sustainable U.S. supply chain could serve as a regional source for nearby emerging markets in the Americas and beyond, where an increasing share of the potential future global wind growth may take place [11].

Action 3.2: Develop Transportation, Construction, and Installation Solutions

Transporting, constructing, and installing the advanced technology described in the *Wind Vision* will present many challenges. Identifying these challenges and developing and implementing solutions to enable deployment of larger wind turbines on taller towers in new regions of the country and offshore will require strong cooperation between industry and government agencies.

As components grow in size and weight, the limitations of ground transport from factory to installation site increase, especially for land-based systems [12]. Issues include safety, the integrity of the public infrastructure, and increased cost of components designed around transportation constraints rather than performance optimization. Industry and state

and local government agencies will need to assess the primary issues and write best practices to support improved logistics planning and clarify the transportation constraints. This will enable original equipment manufacturers and transportation and logistics companies to develop new component designs and logistics solutions to ensure larger turbines can be deployed cost-effectively. The use of larger turbines on taller towers is also affected by Federal Aviation Administration rules, which require additional review for structures over 500 feet. Actions to address these issues are discussed in Section 4.6, *Wind Siting and Permitting*.

ACTION 3.2: Develop Transportation, Construction, and Installation Solutions	
Develop transportation, construction, and installation solutions for deployment of next-generation, larger wind turbines.	
DELIVERABLE	**IMPACT**
Transportation, construction, and installation technology and methods capable of deploying next-generation land-based and offshore wind.	Reduced installed capital costs and deployment of cost-effective wind technology in more regions of the country.

Key Themes: Reduce Wind Costs; Expand Developable Areas
Markets Addressed: Land, Offshore

New construction and installation techniques, materials, and equipment, such as multi-crane lifts of heavy nacelles or concrete towers, will also be needed to install next-generation wind plant technologies. Concepts such as on-site manufacturing and assembly of towers or other components could mitigate some of the challenges presented by larger, heavier components, but will need to be demonstrated before being widely deployed. Dedicated technology demonstration sites could provide a venue to certify new construction and installation technologies without creating risk or otherwise affecting the financing of commercial projects. Once proven, these solutions could then be deployed on a broader scale.

Offshore wind is subject to unique challenges and will require new supply chain investments, including infrastructure and logistics networks. These issues are more thoroughly addressed in Action 3.3.

Action 3.3: Develop Offshore Wind Manufacturing and Supply Chain

Advancement of the U.S. offshore sector toward deployment—with initial projects readying for construction and lease sales establishing a flow of projects—brings into sharper focus issues related to manufacturing capacity, skilled workforce, and maritime infrastructure requirements. Studies commissioned by DOE and released in 2013–2014 provide a knowledge base for considering strategic approaches to planning, promoting, and investing in necessary industrial-scale wind assets in a cost-effective, efficient manner. These studies address port readiness [13][4]; manufacturing, supply chain, and workforce [14]; and vessel needs [15], each under a variety of deployment assumptions through 2030.

ACTION 3.3: Develop Offshore Wind Manufacturing and Supply Chain	
Establish a domestic offshore manufacturing, supply chain, and port infrastructure.	
DELIVERABLE	**IMPACT**
Increased domestic supply of offshore wind components and labor.	Increased economic growth in major offshore ports and regional manufacturing centers.
Key Themes: Increase Economic Value for the Nation **Markets Addressed:** Offshore	

Taken together, these assessments illustrate several principles:

- There is a wide range of economic development and job creation opportunities associated with offshore wind development. The United States has significant existing assets currently at the service of other industries or underutilized that can be deployed in support of offshore wind development.

- While offshore and land-based wind development share some commonalities, the offshore sector—due to larger component size, maritime logistics, rapidly advancing technology, and its early development phase—will require new skills and infrastructure development, and offers new and different economic development opportunities and challenges.

- Industrial infrastructure is primarily a function of market demand. The project-by-project approach of supply chain mobilization that is necessary for the first offshore wind projects will not be an effective or efficient process in planning for industry-scale deployment.

- The scale of deployment needed to support significant private sector investment in new manufacturing facilities, port improvement, and purpose-built vessel construction will likely be associated with regional markets for offshore wind and policies of multiple states.

- Development of the manufacturing base, workforce, and maritime infrastructure necessary to support a viable offshore wind industry will require integrated public and private sector vision, commitment, and investment.

European experience illustrates the significant effect on project cost and risk management that can result from supply chain gaps and vessel shortages [16, 17], and the dangers of losing economic development advantage in the competitive global offshore wind market due to a lack of strategic investment and planning [18]. Supply chain efficiencies have been targeted in the United Kingdom as a key opportunity for lowering the cost of offshore wind power [17]. The United States can capitalize on the lessons from Europe's experience to position the nation to realize offshore wind power's full economic development potential across participating states.

---

4. In addition to the port readiness report, DNV-GL created a port readiness tool (*http://www.offshorewindportreadiness.com*).

4.4 Wind Power Performance, Reliability, and Safety

Wind power is becoming a mainstream, widespread technology. With this progress, asset owner/operators, utilities, and the public expect wind plants to meet the same operational reliability as conventional generation sources. While substantial progress has been made in reliability and availability of systems, significant reductions in overall cost of energy can still be realized through better O&M practices. This is especially true in the offshore environment, where maintenance costs are significantly higher due to more difficult access. **These practices can be accomplished by the activities described in this section:**

- **Improve reliability and increase service life through the development of new technology and better understanding of operational environments;**

- **Develop a world-class database on wind plant operation under normal operating conditions;**

- **Develop understanding of reliability under severe operating conditions;**

- **Develop and document best practices in O&M procedures;**

- **Develop aftermarket technology upgrades and best practices for repowering and decommissioning**

Detailed activities and suggested timelines for action are identified for each of these key areas in Appendix M.4.

Action 4.1: Improve Reliability and Increase Service Life

Reliability and a long economic service life are essential requirements for all power generation systems, and these attributes will become increasingly important for wind power systems as they supply greater portions of U.S. electricity demand. Unplanned replacement of wind turbine components is a major cost to wind plant owners and operators, both in terms of the cost to replace the components and in lost revenue from machine downtime.

ACTION 4.1: Improve Reliability and Increase Service Life	
Increase reliability by reducing unplanned maintenance through better design and testing of components, and through the adoption of condition monitoring systems and maintenance.	
DELIVERABLE	**IMPACT**
Reduced uncertainty in component reliability, and increased economic and service lifetimes.	Lower operational costs and financing rates. Increased energy capture and investment return.

Key Themes: Reduce Wind Costs
Markets Addressed: Land, Offshore

Improving wind turbine component, sub-system, and system reliability can reduce costs for O&M and component replacement, reduce downtime, and potentially reduce wind plant financing costs. Increasing the economic service life of wind power systems from the present 20–24 year design life to 25–35 years can further and significantly reduce the cost of these long-term energy production assets [19].

Strategies to improve reliability and service lifetime can target all phases of the wind power life cycle, including design, testing, manufacturing, operations, maintenance, refurbishment and upgrades, and recycling. Changing the design and certification philosophy at the design phase of product development to include a specific reliability basis will be an essential step. This includes planning for refurbishment, replacement, and product improvement upgrades early in the product development process.

Data from both field and controlled reliability testing are required to inform improved design practices and design standards. Collection and analysis of field data improve the understanding of environments and actual operating conditions in which each component operates and the specific mechanisms that cause early failure. Of particular importance is better understanding of the complex loads imposed in the interior of a wind plant. These data can be supplied by activities discussed in Section 4.1, *Wind Power Resources and Site Characterization*.

In addition to field tests, more comprehensive testing of components and subsystems will inform improved in-service reliability. Interactions between subsystems are a common source of reliability issues—for example, the interaction between a blade pitch bearing and the rotor hub that supports it, which is not perfectly rigid. Accelerated lifecycle testing of critical components under controlled conditions can be used to simulate operating environments. Present blade testing requirements can be augmented to also address testing of the drivetrain, electric power conversion system, and other key subsystems such as the blade pitch control system.

Beyond design for reliability and testing to identify failure modes and validate improved designs, condition monitoring systems can ensure that maintenance actions occur before failures occur. While industry is beginning to transition from traditional time-based component replacement schemes to condition-based component maintenance and replacement, and condition monitoring system sensors are becoming less costly, significant effort is still required to develop predictive analysis methodologies that convert the raw sensor data into actionable maintenance alerts.

Stakeholders can work together to conduct these activities through a cycle of robust design practices, testing and data collection, and targeted research projects, all informing the improvement of reliability standards and design testing.

Action 4.2: Develop a World-Class Database on Wind Plant Operation Under Normal Operating Conditions

Performance and reliability data from wind plants across the country are essential to understanding the state of the current fleet and benchmarking technology improvements. A trusted database will allow research funds to be directed towards the best opportunities to reduce cost of energy while also reducing uncertainty for financiers and insurance providers. Design standards will be improved by better understanding of the operating environment, and wind plant O&M practices will be optimized based on these data. Realizing this database will

ACTION 4.2: Develop a World-Class Database on Wind Plant Operation under Normal Operating Conditions	
Collect wind turbine performance and reliability data from wind plants to improve energy production and reliability under normal operating conditions.	
DELIVERABLE	**IMPACT**
Database of wind turbine performance and reliability data representing the U.S. fleet.	Lower unplanned maintenance costs, lower financing and insurance rates, and increased energy production.

Key Themes: Reduce Wind Costs
Markets Addressed: Land, Offshore, Distributed

require cooperation among industry and research institutions such that maintenance records and turbine controller data can be collected from original equipment manufacturers, owner/operators, and third party service providers, in both warranty and out-of-warranty periods.

Action 4.3: Ensure Reliable Operation in Severe Operating Environments

As wind power installations increased in the United States, manufacturers encountered severe operating conditions such as lightning and erosion, especially in the western plains. With more wind plants being developed in colder climates and offshore installations pending, improvements are needed in turbine design to mitigate the effects of icing, salt water corrosion, and hurricanes. Despite these obstacles, reliable operation needs to be achieved to give financiers and regulators confidence in wind technology. This will require substantial work in data collection and model development, ultimately leading to the improvement of existing standards as well as possible creation of new ones. Research institutions need to collaborate with turbine manufacturers to perform targeted studies of each of these issues. The knowledge gained will facilitate structural and material design improvements to turbine components by the original equipment manufacturers, targeted mitigation solutions from third-party suppliers, and improved O&M practices from service companies.

ACTION 4.3: Ensure Reliable Operation in Severe Operating Environments	
Collect data, develop testing methods, and improve standards to ensure reliability under severe operating conditions including cold weather climates and areas prone to high force winds.	
DELIVERABLE	**IMPACT**
High availability and low component failure rates in all operating environments.	Lower unplanned maintenance costs, lower financing and insurance rates, and increased energy production.
Key Themes: Reduce Wind Costs **Markets Addressed:** Land, Offshore, Distributed	

Action 4.4: Develop and Document Best Practices in Wind O&M

Development of industry best practices is critical to the training and education of the workforce, the safety and efficiency of the work performed, and the energy production of the wind plant.

ACTION 4.4: Develop and Document Best Practices in Wind O&M	
Develop and promote best practices in O&M strategies and procedures for safe, optimized operations at wind plants.	
DELIVERABLE	**IMPACT**
Regular updates to the American Wind Energy Association's O&M Recommended Practices document and other industry-wide documents.	Consistency and improvement of O&M practices and transferability of worker skills.
Key Themes: Reduce Wind Costs **Markets Addressed:** Land, Offshore	

While the American Wind Energy Association's O&M Recommended Practices document [20] is a step in this direction, further improvement is needed to achieve the same level of standardization as conventional power sources. Progress in this activity will require the cooperation of trade organizations, government agencies, and wind industry members over the coming decades, and will necessitate extensive data collection. The result is expected to facilitate improved wind plant energy production, while minimizing integration and environmental impacts and increasing worker safety.

Action 4.5: Develop Aftermarket Technology Upgrades and Best Practices for Repowering and Decommissioning

The market in upgrades from both original equipment manufacturers and third-party suppliers is thriving. Owners of existing wind turbines—which are expected to remain in operation for 20 years or more—will want access to increased energy production, improved reliability, and decreased costs offered by improved technology as it is introduced into the market on new turbines. Rather than choosing more costly complete replacement of existing turbines, industry can continue to devise options for upgrades or refurbishment with replacement components offering the new technology. Specific actions include developing trusted remanufacturing and reconditioning techniques for expensive components; developing improved control systems and using technology from new turbines to accompany retrofits through better operational environment monitoring; and developing component retrofit and upgrade pathways such as larger or better performing rotors and more reliable drivetrains.

Wind turbine owners must decide what to do when wind turbines reach the end of their planned operating life. Options include repowering or refurbishing existing equipment or decommissioning the turbines.

Develop aftermarket upgrades to existing wind
plants and establish a body of knowledge
and research on best practices for wind plant
repowering and decommissioning.

DELIVERABLE	IMPACT
Aftermarket hardware and software upgrades to improve operational reliability and energy capture, along with reports and analyses on wind repowering and decommissioning.	Increased energy production and improved decision-making for aging wind plant assets, including repowering to avoid greenfield development costs.

Key Themes: Reduce Wind Costs
Markets Addressed: Land

Establishing a body of knowledge, best practices,
and strategy on wind repowering and wind plant
retirements and decommissioning can help owners
make cost-effective decisions. Creating such a body
of knowledge requires research, data gathering, and
review of existing practices.

Related actions that can be undertaken by wind
stakeholders include:

- Analyzing and building on California and European
experiences and practices in wind plant repowering
and decommissioning of the earliest installed wind
plants;
- Documenting repowering and decommissioning
practices in other energy, transportation, and
aerospace technologies;
- Developing and refining broadly accepted standards for recertification and life extension of wind
plants and components; and
- Distilling best practices for wind plant
decommissioning.

Success in this activity will require close collaboration
between the wind plant owners and operators who
will provide operational experience and the market
for upgrades; original equipment manufacturers; and
third-party equipment manufacturers who supply
equipment for this market. Stakeholders also need to
work together to find solutions for aging turbines that
satisfy community concerns such as viewscape, land
usage, and other environmental aspects, as well as the
economic concerns of equipment and land owners.

4.5 Wind Electricity Delivery and Integration

Successfully addressing power system integration
issues, while still maintaining electric power system
reliability, is critical to achieving high wind penetrations at reasonable costs. Key issues in this area relate
to increased variability and uncertainty posed by wind
power at various time scales. Methods for managing the power system with moderate-to-high wind
penetrations have evolved, and will likely continue
to evolve as more actual experience is gained with
wind power plants. Utilization of wind forecasting in
operational practice of power systems and advanced
controls on wind turbines can help operators
decide on appropriate reserve levels. In some cases,
operators will be able to deploy wind turbine and
wind plant response capabilities to help manage the
power system. Experience and research demonstrate
these approaches can be executed at reasonable
cost if appropriate actions are taken. If integration
techniques are not appropriate, however, operating
costs of the power system could be too high and wind
deployment impeded.

Aggregate power system generation needs to match
aggregate power system load instantaneously and
continuously. Load, renewable generation such as
wind and solar, and conventional generation all
contribute variability and uncertainty.

Operating the power system with high penetrations of wind power while maintaining reliability at minimum cost requires actions in at least six key areas:

- **Encourage sufficient transmission to deliver potentially remote generation to load and provide for economically efficient[5] operation of the bulk power system over broad geographic and electrical regions;**
- **Encourage the availability of sufficient operational flexibility;**
- **Inform the design of proper incentives for investment in and deployment of the needed flexible resources[6];**
- **Provide advanced controls for grid integration;**
- **Develop optimized offshore wind grid architecture and integration strategies; and**
- **Improve distributed wind grid integration.**

Transmission network design to accommodate large amounts of wind power, which may be developed in remote locations including the Great Plains and the U.S. Continental Shelf, presents challenges. Linking large electrical and geographic areas, however, can help promote reliability and cost-effective bulk system operation [21]. Benefit and cost analyses of new transmission designs are needed to determine whether a given design is promising, and whether AC-only or AC-DC hybrid options make sense. The latter can tie asynchronous AC systems together and deliver wind energy and reliability benefits over large areas.

At high wind power penetrations, maintaining system balance while minimizing wind power curtailment requires that non-wind generators can be operated flexibly. This means generators may need to be ramped (changing output levels) and start/stop more quickly than was done without wind. Many older generators were not designed for the level of flexible operation that would likely be required by the *Wind Vision Study Scenario*. The supply of flexible resources, including demand response and storage as well as flexible conventional generation, needs to be increased to accommodate high levels of wind power. When cost effective, new storage technologies can be considered in the future.

The current electricity industry structure comprises regulated markets. The precise form of these markets varies, from regulated monopolies in much of the West and Southeast to regional transmission operator/independent system operator markets in much of the East, Texas, and California. A prerequisite to ensure that sufficient operational flexibility is available in real-time is an accurate assessment of future flexibility needs, along with market incentives to develop this level of flexibility. It is thus necessary to develop and implement operating practices and market structures that result in cost-effective power system operation, while maintaining reliability of delivery with high levels of wind power. These operating practices and market structures will inform the design of incentives to develop and deploy flexible resources as they are needed. As one example, specifications for new natural gas-fired capacity could require substantial operational flexibility.

Developing transmission, flexible generation, and market incentives are functions of the power system industry, including utilities, regional transmission operators, the regulatory community, and other entities involved in delivery of electricity. Market incentives are primarily market structures and designs that encourage flexibility and are part of the bulk power system. When these incentives are not in place, there can be a decline in flexibility. An example is the decline in frequency response in the Eastern Interconnection, caused in large part by a lack of market incentive [22, 23]. There is an important role for stakeholders in helping to develop best practices in power system operation and design, as well as in designing both physical and institutional systems to support achieving the *Wind Vision*. It will be critical to disseminate that information to power system operators and to support implementation of best practices.

Detailed activities and suggested timelines for action are identified for each of these key areas in Appendix M.5.

---

5. "Economically efficient" denotes the most cost-effective way of achieving the goal of operating the power system reliably with a given level of wind power. An outcome is economically inefficient if it provides the same level of reliability at higher cost.

6. For the purposes of this discussion, the term "resources" includes flexible generation, potential demand response, and appropriate storage.

Table 4–2. Texas Installed Wind Capacity and ERCOT Curtailment during CREZ Transmission Consideration, Approval, and Construction (2007–2013)

	2007	2008	2009	2010	2011	2012	2013
Texas installed wind capacity (MW)	4,446	7,118	9,410	10,089	10,394	12,214	12,354
Curtailment in ERCOT (fraction of potential wind generation)	1.2%	8.4%	17.1%	7.7%	8.5%	3.8%	1.2%

Note: The CREZ transmission project was approved by the Public Utility Commission in 2008. Construction was completed in 2013. The great majority of Texas' wind capacity is located in the ERCOT region (89% at the end of 2013).

Source: DOE 2008–2014 [25, 26, 27, 28, 29, 30, 31]

Action 5.1: Encourage Sufficient Transmission

Transmission is required to move wind energy from wind-rich regions to load centers. Balancing over large areas also requires transmission and can reduce operating cost. Studies are necessary to develop alternative transmission network designs that balance a range of technical, economic, and regulatory issues. While transmission expansion slowed for many years

---

ACTION 5.1: Encourage Sufficient Transmission

Collaborate with the electric power sector to encourage sufficient transmission to deliver potentially remote generation to electricity consumers and provide for economically efficient operation of the bulk power system over broad geographic and electrical regions.

DELIVERABLE	IMPACT
Studies, methodologies, and validated tools that inform cost-effective, reliable electricity delivery from wind power and all other generation types.	Increased transmission, reduced electricity costs, and increased wind generation with less curtailment.

Key Themes: Reduce Wind Costs; Expand Developable Areas
Markets Addressed: Land, Offshore

---

in the United States, the Electric Reliability Council of Texas's (ERCOT's) development of the Competitive Renewable Energy Zone (CREZ) transmission build-out demonstrates that the issue can be addressed. The CREZ project was enabled by the Texas Legislature in 2005 and Public Utility Commission action in 2008. Now complete, the 3,588 miles of Competitive Renewable Energy Zone transmission lines carry approximately 18,500 MW of wind power from West Texas and the Texas Panhandle to load centers in Austin, Dallas-Fort Worth, and San Antonio [24]. The CREZ line additions have substantially reduced wind curtailment in the ERCOT region, as discussed in Chapter 2 and summarized in Table 4-2.

Other regions are following Texas's lead in adopting practices to enable long-needed grid upgrades that will benefit consumers while also reducing wind curtailment and enabling new wind development. The Midcontinent Independent System Operator has adopted similar broad cost allocation practices for a set of transmission lines, called the Multi-Value Projects, which will potentially integrate nearly 15 GW of new wind capacity. The Southwest Power Pool has similarly adopted a highway/byway transmission cost allocation policy and is making progress towards building a set of wind-serving lines called the Priority Projects. PJM Interconnection's State Agreement Approach allows projects with public policy benefits to be constructed when states agree to fund them, similar to the Texas CREZ.[7] In addition, PJM members

---

7. PJM Interconnection is a Regional Transmission Organization within the Eastern Interconnection. See *http://www.pjm.com/about-pjm.aspx* for more information.

have approved a new category of transmission projects called Multi-Driver Projects. These projects, which are subject to approval by the Federal Energy Regulatory Commission, would allow transmission upgrades with multiple benefits to proceed where they otherwise might not. Some other regions of the country have initiated coordinated planning activities but still lack the transmission cost allocation and planning practices essential for enabling multi-state transmission investment.

Transmission expansion is difficult but vital, because it spans issues ranging from detailed technical stability analysis to broad concerns about regulatory cost allocation. Complexity is further increased by the fact that transmission inherently connects large geographic areas. This raises the number of stakeholders and regulatory jurisdictions, creating the potential for multiple interveners in the approval process. Transmission projects also have very long lifecycles, which enhances their economic benefits but increases uncertainty in the value analysis. Transmission will not only help to effectively integrate wind power, but also increase bulk system reliability and reduce operating costs for the existing power system. This can provide benefits for the electricity ratepayer, but it complicates both the analysis and regulatory treatment of transmission expansion. For further discussion of transmission benefits analysis, see (for example) Chang et al, 2013 [32].

Transmission investment is also "lumpy," meaning that it is typically not cost-effective to build low-voltage lines at lower cost that may need upgrading in the lifetime of those lines (often 50 or more years). This implies greater levels of uncertainty surrounding the useful life of transmission and can suggest that transmission investments be made to accommodate distant-future needs and cover broad geographic regions. This may imply the need and opportunity for wind stakeholders to collaborate with others to inform large-scale, inter-regional, long-term planning to capture the economic benefits.

Several long distance DC transmission lines were under consideration as of 2014. Benefits of adding these lines include delivering remote wind and solar generation to load centers, improved reliability, reduced regulation and spinning reserve requirements, increased generator availability, and optimized generation dispatch benefits that capture diversity throughout the footprint.

To achieve a transmission infrastructure of the type that would support the *Wind Vision Study Scenario*, complex rules regarding transmission build-out over multiple jurisdictions will need to be addressed. The *Wind Vision* analysis finds the new transmission requirement in the *Wind Vision Central Study Scenario* is 2.7 times greater than in the *Baseline Scenario* by 2030, and 4.2 times greater by 2050 (see Chapter 3 for more detail).

There is evidence that initiatives such as the Texas CREZ can achieve this objective on an intra-state and intra-jurisdictional basis. While potentially achievable, however, transmission that crosses state boundaries may be difficult on such a large scale because local concerns may not align with broader social benefit.

Other issues that need to be addressed include determining whether system dynamics and system inertia[8] will be affected by large penetrations of wind power, and, if so, what cost-effective mitigation approaches can be used. Methods to analyze these impacts are not mature, and therefore need to be developed and refined. Once new methods and tools are developed, they need to be tested and then applied to expected high wind power penetrations at specific locations on the bulk power system to determine the potential impacts and mitigation strategies, if needed. There are new and emerging advanced control technologies that may be helpful, and these also need to be more fully developed and tested.

Action 5.2: Increase Flexible Resource Supply
Wind generation increases both the variability and uncertainty of the aggregate power system and, through displacement, reduces the amount of conventional generation that is under system operator control and available to balance net load. Sub-hourly energy markets and larger balancing areas reduce balancing requirements, but increasing the resource pool from which to balance is still necessary to cost-effectively integrate wind power. More flexible

---

8. System inertia is a measure of the ability of the system to ride through short-term disturbances by drawing on the mechanical "flywheel" inertia of spinning power plant rotors.

resources,[9] along with more flexible operating practice in the power system industry, are needed to integrate large amounts of wind power. Simply increasing the supply of flexible resources is a necessary, but not sufficient, condition to achieve flexibility in power system operations. In order for flexibility targets to be achieved, operating and market rules must not hinder access to the physical flexibility in the ground. Otherwise, physical flexibility can be stranded and thus unavailable to the power system operator.

ACTION 5.2: Increase Flexible Resource Supply

Collaborate with the electric power sector to promote increased flexibility from all resources including conventional generation, demand response, wind and solar generation, and storage.

DELIVERABLE	IMPACT
Analysis of flexibility requirements and capabilities of various resources. Frequent assessments of supply curve for flexibility. Implementation of cost-effective rules and technologies.	Reduced wind integration costs, reduced wind curtailment, improved power system efficiency and reliability.

Key Themes: Reduce Wind Costs; Expand Developable Areas
Markets Addressed: Land, Offshore, Distributed

Because of the complexity of the power system and the uncertainty surrounding specific locations of new generation and transmission, analysis activities can help quantify the value of flexible resources. These resources include (but may not be limited to) reciprocating engine-driven generators, advanced aero-derivative combustion turbines, flexible combined cycle generators, demand response, purpose-built storage (e.g., pumped hydroelectric storage or large batteries) and inherent storage (e.g., domestic water heaters or plug-in automobiles with charge-discharge capability). Expanding the functionality of demand response and inherent storage provides opportunities for stakeholder action, including:

- Developing a more comprehensive assessment of these resources, including industrial loads;
- Reducing the cost of implementing demand response through development of appropriate monitoring and control technologies;
- Expanding the range of services provided by demand response, including frequency response, voltage support, and congestion relief;
- Analyzing the effect of new, innovative market designs, such as the influence of performance-based rates for frequency regulation per Federal Energy Regulatory Commission Orders 755 and 784, the impact of intra-hour scheduling requirements per Federal Energy Regulatory Commission Order 764, and the role of scarcity pricing and the intersection with capacity markets; and
- Coordinating wind with hydro and solar to complement natural gas ramping to expand flexible resource supply and demand response capabilities.

There has been significant progress in developing flexible generation. For example, reciprocating engine plants can start within 60 seconds and fully load in 5 minutes, providing value for regions with high wind and solar penetration. There is no limit to the number of starts for these units, and no cycling cost. Coupled with simple cycle heat rates of approximately 8,800 BTU per kilowatt-hour (42% efficiency), plants such as these provide both flexibility and efficiency. A challenge remains in assuring that flexibility is correctly valued with appropriate reliability rules, operating practices, and bulk power system market design incentives.

Opportunities for stakeholder engagement and other collaborative efforts go beyond analysis of benefits and development of optimal utilization strategies. Technology-neutral reliability rules, operating practices, and market incentives can prescribe the required physical characteristics for flexible resources. This technical neutrality fosters competition between technologies and allows for advancements that may result in new sources of flexibility unforeseen at the time of rule development. One prerequisite in achieving a flexible power system is the creation of incentives that foster the development of needed resources.

---

9. Flexibility of resources, which can be either generation or flexible demand or storage, is generally defined as the ability to change states quickly. Thus fast ramping and short start-up, shut-down, and up-/down-times are measures of flexibility.

Action 5.3: Encourage Cost-Effective Power System Operation with High Wind Penetration

Increasing industry understanding of wind integration and developing appropriate operating practices and technology-neutral market rules are necessary to further realize how to economically maintain power system reliability while accommodating increasing amounts of wind generation.

It is also necessary to inform power system operators which practices work and which do not by disseminating findings via publications, workshops, and conferences. This activity provides the scientific background necessary to help promulgate operating best practices, such as sub-hourly energy scheduling and balancing over larger areas, which have the potential to significantly reduce wind integration costs. This activity also illustrates the need for more flexible resources such as fast-starting conventional generation and increased demand response, which can also substantially reduce wind integration costs.

ACTION 5.3: Encourage Cost-Effective Power System Operation with High Wind Penetration

Collaborate with the electric power sector to encourage operating practices and market structures that increase cost-effectiveness of power system operation with high levels of wind power.

DELIVERABLE	IMPACT
Coordination of wind integration studies at the state and federal levels and promulgation of practical findings, especially to entities with less wind integration experience.	Increased wind integration levels, appropriate amounts of operating reserves, reduced curtailment, lower integration costs.

Key Themes: Reduce Wind Costs; Expand Developable Areas
Markets Addressed: Land, Offshore

Graph Removed Due to Copyright Restrictions

Figure 4–1. Increased balancing area size and faster scheduling reduce regulation requirements.

Larger balancing areas and faster generation dispatch (sub-hourly energy markets) considerably reduce wind integration costs. Figure 4-1 demonstrates that requirements for regulation—a relatively expensive balancing service—are reduced substantially as balancing area size is increased and the dispatch interval is decreased. For example, the regulation requirement for a large balancing area drops from 4,000 MW to just over 1,000 MW as the dispatch timeframe drops from 60 minutes to 10 minutes, and the forecast lead time is reduced from 40 minutes to 10 minutes. Analysis may be required to quantify benefits in regions that are not already implementing sub-hourly energy scheduling or that operate with small balancing areas. Implementation of best practices should also be supported.

The electric sector, with the assistance of DOE, its national laboratories, and federal and state regulators, support development of advanced techniques to reduce wind integration costs as well as studies that quantify the effects of potential regulatory and market structures. Such techniques and studies should seek to accurately encompass multiple balancing areas and regions as well as help promulgate best practices, such as optimization of flexibility reserve. These advanced methods can be used to address technology neutrality concerns, assuring that all technologies are treated equally in reliability rules and market structures.

Action 5.4: Provide Advanced Controls for Grid Integration

The bulk power system needs several ancillary services to help provide reliability and balancing capability. Wind turbines are being developed that can help with voltage control, regulation (automatic generation control), synthetic inertial response, and frequency regulation. Some of these features are untested, and, in many parts of the United States, wind turbine owners and operators have no incentive to provide these services because no market mechanism exists to pay the owners for providing these added capabilities. There is also a need to provide controls at the wind plant level, which would allow wind plants to behave more like conventional generation. The wind stakeholder community can collaborate with others to develop needed control strategies at the wind plant level, building upon newly emerging turbine capabilities.

ACTION 5.4: Provide Advanced Controls for Grid Integration

Optimize wind power plant equipment and control strategies to facilitate integration into the electric power system, and provide balancing services such as regulation and voltage control.

DELIVERABLE	IMPACT
Advanced wind turbine and wind plant controls that can be used to provide voltage support, regulation, synthetic inertial response, and frequency regulation by wind plants. Bulk power market designs and/or tariffs are necessary to pay for these services.	Allows power system operator access to additional flexibility from wind plants, when it is economical or necessary for reliability. This will reduce cost and increase reliability.

Key Themes: Reduce Wind Costs; Expand Developable Areas
Markets Addressed: Land; Offshore

Action 5.5: Develop Optimized Offshore Wind Grid Architecture and Integration Strategies

In most cases, offshore wind power plants will be constructed in waters near large urban load centers. The *Wind Vision Study Scenario* includes the construction and integration of multiple offshore wind plants. Each project is individually responsible for the interconnection that brings power to shore. These power delivery systems will be built on public waterways and connected to the on-shore grid infrastructure. Under this activity, aggregating the power export systems for multiple offshore facilities is expected to lower the cost of offshore transmission and minimize impacts to coastal ecosystems where cables are routed. Several strategies are under consideration in the United States to develop optimized architectures for the orderly construction of an offshore grid. As part of this effort, close coordination between state and federal agencies is needed to streamline the offshore permitting process and reduce regulatory uncertainty.

Offshore wind electricity will typically be injected into heavily congested urban centers. As such, the integration of offshore wind in certain markets will have global utility effects that reduce the market price of electricity, at least for the near term. The capacity value of offshore wind differs from that of land-based wind and, in some regions, provides stronger matching with load during peak summer months. Both of these effects significantly influence the economics of offshore wind technology for the *Wind Vision Study Scenario*.

Action 5.6: Improve Distributed Wind Grid Integration

While utilities generally have experience integrating wind into the grid as well as confidence in land-based wind systems to deliver reliable power, distributed wind faces challenges in gaining a similar level of confidence and integration experience. The grid effects of distributed wind generation, alone and integrated with other forms of distributed generation, need to be better understood in order to facilitate mitigation and removal of integration barriers and to accelerate deployment. Better distribution system modeling tools, informed utilities, and standards development can reduce costs and increase confidence in distributed wind integration. This will improve prospects for increased distributed wind deployment. As an example, a new revision of IEEE 1547 Standard for Interconnecting Distributed Resources with Electric Power Systems [34] is underway as of 2014. This revision will establish a framework for distributed generation that supports the grid and allows high levels of penetration.

4.6 Wind Siting and Permitting

As with any form of energy, there are impacts to the natural surroundings associated with wind power. Wind is a comparatively clean source of energy with many positive attributes, such as no emissions, no air or water pollution, and no use of water in the generation of electricity. If improperly sited, however, wind power facilities may present a number of socioeconomic, conflicting use, and environmental risks. Care needs to be taken in the siting of wind power facilities to ensure the potential for negative impacts from construction and operation is minimized to the greatest extent practicable. These risks, or even the perception of risk, may pose obstacles to wind deployment throughout the United States. Regulators and other energy-sector decision makers need to ensure that energy generation choices reflect the public interest. To address this need, actions in this section focus on the real or perceived undesirable impacts of wind power and the development of regulations and policies that support wind development while equitably minimizing its real and perceived impacts.

Some potential impacts of wind are well-known and can be reduced and mitigated through existing siting and permitting processes (see Chapter 2 for more details). Other potential issues demand more research, either because the actual impacts are not quantifiable or because particular impacts to ecosystems or species of concern are not well understood. In some cases, there is also limited practical experience upon which decision makers can draw, such as with offshore wind on U.S. coasts or in the Great Lakes, or because of new or developing regulatory frameworks. The cost-effectiveness of new impact reduction and mitigation methods should be taken into account to understand if these methods are viable within the highly competitive U.S. energy sector or even necessary from a practical standpoint, as zero-impact development is not possible. **Five overriding actions important to responsible expansion of wind deployment are discussed in this section. They are:**

- **Evaluate potential competing public use challenges related to wind plants such as radar, aviation, land use, residential impact, commercial fisheries, maritime shipping, and navigation;**

- **Develop and disseminate relevant information on siting and mitigation strategies for wildlife and other natural resource concerns;**

- **Continue to gather and disseminate accurate information to the public on local impacts of wind power deployment and operations;**

- **Collaborate to inform streamlined regulatory frameworks for wind development on public land, and do so with the understanding that flexibility is needed to manage variability of wind projects by location; and**

- **Develop commonly accepted siting frameworks and assessment tools that can be used to inform faster wind site pre-screening.**

Detailed activities and suggested timelines for action are identified for each of these key areas in Appendix M.6.

Action 6.1: Develop Mitigation Options for Competing Human Use Concerns

Wind power plants often cover the same geographical area as other potential uses, bringing about discussions of conflicts. In most cases, wind technology can operate without impacting other uses, such as with most civil aviation. In some cases, however, such as with military and weather radar systems, the potential interactions need to be better understood and may be location- and use-specific. In cases such as navigation, military operations, and commercial and recreational fisheries, detailed discussions with potentially affected stakeholders are needed. Other potential impacts, such as those on local viewsheds or tourism, are often a matter of public perception. Addressing these may require engagement with a broader range of stakeholders. To effectively characterize the challenges and develop mitigation options for any of these issues, detailed discussions and—in many cases—experimental research will be required.

ACTION 6.1: Develop Mitigation Options for Competing Human Use Concerns

Develop impact reduction and mitigation options for competing human use concerns such as radar, aviation, maritime shipping, and navigation.

DELIVERABLE	IMPACT
A better understanding of the impacts of wind development and appropriate mitigation options leading to streamlined site assessment and trusted hardware and software technology solutions that address the most pressing competing use conflicts.	Decreased impact of all wind technologies allowing project developers to site wind projects while limiting competing public use impacts.

Key Themes: Reduce Wind Costs; Expand Developable Areas
Markets Addressed: Land, Offshore, Distributed

A large number of key conflicting uses are already understood for land-based wind development. Competing uses for expanded offshore wind development are less defined, but aviation safety, navigation safety, radar, and competing economic uses are known to be of importance. One of the initial steps for ensuring thorough understanding of competing uses is the development of expanded geographic information tools in which multiple data sets related to land and water uses are collected from a broad group of public and private stakeholders. A common, vetted, and complete database will help facilitate discussions and planning between wind development and other human use concerns. For competing uses that are a matter of safety, security, or similar concerns, detailed understanding of the potential conflicts needs to be developed with participation of all concerned stakeholders. With this understanding, mitigation strategies (which may include new hardware and software technologies) can be developed, tested, and verified to reduce impacts and enable cost-effective wind deployment that meets stakeholder needs.

Action 6.2: Develop Strategies to Minimize and Mitigate Siting and Environmental Impacts

Potential impacts of wind deployment on wildlife and other ecological systems include the direct mortality of individual birds and bats; injury or behavioral impacts to marine life as a consequence of construction or operational noise in the offshore space; and fragmentation or disturbance of wildlife habitat. Although understanding already exists about these impacts, filling knowledge gaps will require nation-wide investment in species-specific, long-term research. Such research has historically fallen to individual project developers, resulting in a patchwork of sometimes inconsistent research that makes reaching a national consensus difficult.

ACTION 6.2: Develop Strategies to Minimize and Mitigate Siting and Environmental Impacts

Develop and disseminate relevant information as well as minimization and mitigation strategies to reduce the environmental impacts of wind power plants, including impacts on wildlife.

DELIVERABLE	IMPACT
Accurate information and peer-reviewed studies on actual environmental impacts of wind power deployment, including on wildlife and wildlife habitat.	Decreased environmental impact by all wind technologies, improved understanding of the relative impact of wind development, defined methodologies to assess potential impacts and risks, and shorter and less expensive project deployment timelines.

Key Themes: Reduce Wind Costs; Expand Developable Areas
Markets Addressed: Land, Offshore, Distributed

Determining whether additional measures need to be employed and what those measures should be requires building from existing understanding of wind power's effects, balancing wind power's positive attributes with its potential negative impacts, and considering these impacts in comparison to other forms of energy generation. Potential mitigation options identified through this process should be developed, tested, evaluated for cost-effectiveness in comparison to the expected benefit, and put into practice as needed. Outreach to a broad range of stakeholders should also continue, to ensure interested parties understand the true and relative impacts of expanded wind deployment. This will permit contextual discussion on relative environmental impacts of wind specifically and the power sector in general, reducing the chance that disproportionately burdensome requirements will be implemented for wind. As turbine development moves into new areas and the effects of climate change become more pronounced, the impact of wind development and the status of specific species may change. Ongoing assessments and research will be required.

In order to understand the potential long-term impacts of expanded wind development and enhance coexistence of wind power and wildlife, large-scale wildlife and metocean baseline studies will be required. Existing avoidance and minimization options, such as bat deterrent technology and reduction of impacts through operational minimization measures (changes in turbine operations during high risk periods, such as fall migration) also need to be further assessed. This information will help determine effectiveness and appropriate application of these strategies along with other conservation support approaches, such as habitat preservation. The end goal is to provide the industry with multiple, cost-effective ways to reduce—and, to the extent practicable, fully offset—the expected impacts of specific wind projects.

Action 6.3: Develop Information and Strategies to Mitigate the Local Impact of Wind Deployment and Operation

Wind deployment can pose real or perceived public impacts to communities and individuals that live in close proximity to wind power facilities of all sizes. Although wind offers many positive attributes related to the environment (e.g., avoiding air and water pollution, reductions in water usage), as well as to jobs, local land payments, taxes, and other community

benefits, there are also potential challenges such as visual or aesthetic impacts, annoyance associated with turbine noise, and physical safety issues such as ice-throw. Location-specific public opinion can be negatively affected due to misconceptions about these concerns or a lack of understanding of wider community benefits.

ACTION 6.3: Develop Information and Strategies to Mitigate the Local Impact of Wind Deployment and Operation	
Continue to develop and disseminate accurate information to the public on local impacts of wind power deployment and operations.	
DELIVERABLE	**IMPACT**
Accurate information and peer-reviewed studies on the impacts of wind power deployment that can be used and shared through a variety of platforms.	Decreased impact by all wind technologies, defined methodologies to assess potential impact, and shorter and less expensive project deployment timelines.

Key Themes: Expand Developable Areas
Markets Addressed: Land, Offshore, Distributed

As discussed in Chapter 2, substantial information already exists about many of these impacts. In some instances, however, more is still to be understood and documented about the specific impacts and benefits to communities as a result of wind power development. As wind turbines are deployed into new areas or locations in closer proximity to population centers, further research on public impacts will be needed to reduce or eliminate concerns for specific projects and to mitigate real community impacts in an appropriate and cost-effective manner. Stakeholders need to develop a better documented understanding of community concerns and expected benefits, foster accurate assessment tools, and identify appropriate mitigation strategies to address the largest impacts. Ongoing outreach by the wider stakeholder community is needed, so that communities that may be affected by a new wind power development can make decisions based on current, accurate, and widely accepted information.

Action 6.4: Develop Clear and Consistent Regulatory Guidelines for Wind Development

Wind projects trigger a number of regulatory requirements at federal, state, and local levels. The regulations and associated governing bodies that might affect any single wind project depend on a number of variables, including whether a project is on public or private land, the state and locality of the project, and its size. This variation in permitting processes and requirements across locations and government levels can cause inconsistencies in project timelines and increase project risk. In addition, uncertainty about future federal regulatory actions that might affect wind projects is causing hesitation in certain areas of the country, such as those with populations of bat species that may be listed as threatened or endangered in the near future. Effective mitigation measures can help counteract this uncertainty by providing industry with tools to address new regulations and meet permitting requirements.

ACTION 6.4: Develop Clear and Consistent Regulatory Guidelines for Wind Development	
Streamline regulatory guidelines for responsible project development on federal, state, and private lands, as well as in offshore areas.	
DELIVERABLE	**IMPACT**
Defined regulatory guidelines for the deployment of offshore, land-based, and distributed wind turbines, developed in collaboration with the wind industry to provide comprehensible and geographically consistent regulations for the deployment of wind technologies.	Allows developers to clearly understand the processes to deploy wind technologies on federal, state, or private lands, thus reducing costs.

Key Themes: Reduce Wind Costs; Expand Developable Areas
Markets Addressed: Land, Offshore, Distributed

Concise regulatory guidelines are needed that are easy for developers to understand and that address stakeholder needs up front (to avoid conflicts mid-development), such as robust pre-application processes. Guidelines will vary across the country and between levels of government due to jurisdictional, social, and environmental differences. Consistency across agencies and levels of government in such features as the types of information needed to apply for permits, permitting timelines, and opportunities for direct coordination between developers and multiple agencies and levels of government could make permitting easier and more efficient and predictable for developers.

Action 6.5: Develop Wind Site Pre-Screening Tools

Existing requirements, processes, and frameworks for siting wind projects are often loosely coordinated or completely uncoordinated. Such tools range from those to inform site selection and permitting, including tools used to conduct noise or flicker assessments, to those used for initial site screening. Tools can be proprietary, fee-based, or publicly available, and none are housed in a central location or consistently used.

ACTION 6.5: Develop Wind Site Pre-Screening Tools	
Develop commonly accepted standard siting and risk assessment tools allowing rapid pre-screening of potential development sites.	
DELIVERABLE	**IMPACT**
A single or series of interlinked siting tools that support wind turbine siting.	Decreased permitting time while easing permitting processes, leading to lower project development costs with improved siting and public acceptance.

Key Themes: Reduce Wind Costs
Markets Addressed: Land, Offshore, Distributed

Despite the broad range of types and access models for siting tools, no commonly accepted guidelines or set of tool standards exists to ensure such tools are accurate or uniformly applied to inform siting decisions. As a result, organizations on opposing sides of a siting dialogue will often report varying results because they are using different models and assumptions to address similar questions. Additionally, there are no clearly defined screening approaches that allow federal or state regulators to quickly assess potential projects. This results in a more formal and lengthy assessment process, even for projects with limited potential conflicts. While it is impractical to develop universal pre-screening tools that will apply to every situation, there are benefits to providing common best practices where applicable and identifying opportunities to improve efficiencies among federal and state agencies for siting on public lands.

The creation of siting tools should be approached with the understanding that there is broad diversity in wind plant development and informational requirements that are typically based on local and regional concerns. Any guidelines for siting should also be considered conceptual in context. If those conditions are met, the creation of trusted siting tools and wind plant development guidelines can support local wind development while reducing costs through streamlined permitting and the minimization of additional regulatory requirements. These guidelines and accompanying software tool standards are expected to help ensure project assessment accuracy. These activities should help facilitate responsible wind plant development.

4.7 Collaboration, Education, and Outreach

Wind power development has experienced remarkable growth in terms of both deployment and technology innovation. The wind industry is seeing generational changes over the course of years, not decades, which can make it challenging for people not directly involved to stay abreast of this rapidly changing industry. Collaboration among domestic and international wind plant developers and operators, researchers, and other stakeholders during this time of rapid change facilitates learning about new approaches and technical advances that can lead to increased turbine performance, shorter deployment timelines, and lower overall costs.

Public perceptions and regulatory treatment of wind power generation are also influenced by public information that may be incorrect or misleading. Without active collaboration among interested parties, the education of policymakers at all levels, and outreach to stakeholders and the public in general, outdated perceptions of wind power will prevail, limiting the technology's potential and increasing overall project costs.

Given the rapidly evolving wind technology and deployment landscapes, achieving Wind Vision deployment levels will require:

- **Providing information on wind power impacts and benefits to increase public understanding of societal impacts; and**
- **Fostering international exchange and collaboration on technology R&D, standards and certifications, and best practices in deployment.**

Detailed activities and suggested timelines for action are identified for each of these key areas in Appendix M.7.

Action 7.1: Provide Information on Wind Power Impacts and Benefits

Decision makers and the public often lack thorough knowledge about the social costs and benefits of different electricity generation options. As such, decisions are sometimes made about electricity options based on perception, without clear understanding of the actual impacts of those options. These perceptions can influence project permitting and siting timelines, and—if negative—can potentially increase project costs. Accurate, objective, and accessible information about the actual impacts and benefits of wind power can help stakeholders make decisions about wind that are right for their communities.

Quantitative analysis and public dissemination efforts are needed from both public and private sectors of the wind community regarding the relevant positive and negative externalities, including economic outcomes. These efforts need to put potential risks of wind development in the context of the potential benefits, such as jobs, tax revenues for local communities, and avoided environmental impacts. Balanced information will improve decision making about wind development and ensure deployment takes place in an environmentally and socially responsible manner. To the extent possible, impact reduction and mitigation techniques for real impacts on a regional or location-specific basis also need to be articulated. Information should include unique considerations for offshore, land-based, and distributed wind developments.

ACTION 7.1: Provide Information on Wind Power Impacts and Benefits	
Increase public understanding of broader societal impacts of wind power, including economic outcomes; reduced emissions of carbon dioxide, other greenhouse gases, and chemical and particulate pollutants; less water use, and greater energy diversity.	
DELIVERABLE	**IMPACT**
Information and peer-reviewed studies delivered in a stakeholder-targeted method that provides accurate information on the impacts and benefits of wind power independently and in relation to other energy choices.	Retention or expansion of areas open to wind development; decreased fear and misconceptions about wind power; lower project deployment costs and timelines; all leading to more wind installations, better public relations, and lower costs of power.

Key Themes: Expand Developable Areas; Increase Economic Value for the Nation
Markets Addressed: Land, Offshore, Distributed

Action 7.2: Foster International Exchange and Collaboration

The wind industry has become a global trade. Although markets are dominated by Europe, Asia, and the United States, expanded potential exists worldwide. For the wind industry to remain vibrant, an international approach to market development and research collaboration should be considered. Expanding beyond development of wind turbine standards as discussed in Action 2.2, international exchange and collaboration will be required to provide market consistency for U.S. manufacturing and allow global experts to work collaboratively to address ongoing research questions.

International exchanges and expanded information sharing through multilateral organizations such as the International Energy Agency and the International Renewable Energy Agency provide three key benefits: 1) exchange of ideas, research methods, and results among private and public researchers and educational professionals; 2) expanded knowledge of the applicability of wind technology; and 3) experience addressing the deployment challenges of integration, public acceptance, environmental impact, and competing land use. The resulting expansion of wind deployment will allow for increased research and data that can lead to lower costs, and will open export markets for U.S. manufacturing. Along with continued domestic demand, this growing international market can help stabilize the U.S. wind industry and allow industry-wide efficiency improvements.

ACTION 7.2: Foster International Exchange and Collaboration	
Foster international exchange and collaboration on technology R&D, standards and certifications, and best practices in siting, operations, repowering, and decommissioning.	
DELIVERABLE	**IMPACT**
Expanded international collaboration including information sharing, joint research, and staff exchanges allowing expanded education about wind power and expert collaboration from across the wind industry.	Expanded understanding of benefits of wind power across the energy sector; expanded cross-industry collaboration on pressing research topics.

Key Themes: Reduce Wind Costs; Expand Developable Areas
Markets Addressed: Land, Offshore, Distributed

Specific actions for international collaboration include an increased number of cross-border research projects funded by various parties; extended collaboration on the development and use of testing infrastructure; and expanded researcher and academic exchanges—including, for example, permanent researcher-in-residence programs at national laboratories worldwide. Greater international collaboration on the development of wind power research agendas would also be useful.

4.8 Workforce Development

Realizing *Wind Vision Study Scenario* deployment levels and the associated benefits requires a robust and qualified workforce to support the industry throughout the product lifecycle. The industry needs a range of wind professionals, from specialized design engineers to installation and maintenance technicians, to enable the design, installation, operation, and maintenance of wind power systems. To support these needs, advanced planning and coordination are essential to educate a U.S. workforce from primary school through advanced degrees.

Programs at the primary school level introduce developing students to the role of renewable technologies and the range of skills needed to address market requirements. Programs at the secondary school level can add detail and context about wind and other renewable technologies, including practical applications and the scientific and mathematical elements required. Activities targeted at trade workers through community college and vocational technical certification processes supply the wind industry with the much needed technical workforce to install and maintain the expanded fleet of wind plants described in the *Wind Vision*. Finally, specialized skills are developed at the college and advanced degree levels to support wind turbine design, innovation, manufacturing, project development, siting and installation, and additional professional roles.

This section discusses development of comprehensive training, workforce, and educational programs designed to encourage and anticipate the technical and advanced-degree workforce needed by the industry.

Detailed activities and suggested timelines for action are identified for this key area in Appendix M.8.

Action 8.1: Develop Comprehensive Training, Workforce, and Educational Programs

Since wind is a relatively new entrant in domestic and international energy markets, the wind power educational and training infrastructure lags behind that of other major energy technologies. A degree in petroleum engineering is available at a wide range of academic institutions, but similar degrees in wind engineering are only available at a handful of schools.

The absence of common understanding and defined credentials in the land-based, offshore, and distributed wind industries leads to on-the-job training, which increases safety risks for operational staff and leads to errors and inefficiencies. The development of a nationally coordinated educational system addressing all levels will require the collaboration of multiple U.S. federal and state agencies, industry, and the educational community.

ACTION 8.1: Develop Comprehensive Training, Workforce, and Educational Programs	
Develop comprehensive training, workforce, and educational programs, with engagement from primary schools through university degree programs, to encourage and anticipate the technical and advanced-degree workforce needed by the industry.	
DELIVERABLE	**IMPACT**
A highly skilled, national workforce guided by specific training standards and defined job credentials to support the growth of the wind industry.	Sustainable workforce to support the domestic and as appropriate the expanding international wind industry.

Key Themes: Reduce Wind Costs, Increase Economic Value for the Nation
Markets Addressed: Land, Offshore, Distributed

An estimated total of more than 50,000 onsite and supply chain jobs were supported by wind investments nationally as of the end of 2013 [35]. As detailed in Chapter 3, the central sensitivity case of the *Wind Vision* scenario is projected to support a robust domestic wind industry, with jobs from investments in new and operating wind plants ranging from 201,000 to 265,000 in 2020, and increasing to 526,000 to 670,000 in 2050. The *Wind Vision Central Study Scenario* relies on the Energy Information Administration's Annual Energy Outlook 2014 reference data [36] and other literature-derived model inputs and is intended to reflect the central estimate of future effects. The expected expansion of international wind development will greatly increase these needs, further taxing domestic staffing needs but allowing an excellent opportunity for U.S.-based academic institutions interested in renewable energy technology development.

Cross-governmental coordination can help federal and state institutions efficiently mobilize activities in the wind industry. Creating national wind training standards for community college and university sectors requires vision, momentum, and focus in advance of growing demands for skilled individuals; the development of educational programs is a long-term, time-intensive, and expensive process. This foresight will prepare resources to respond to evolving market demands. Stakeholder actions to support the development and implementation of a comprehensive, wind-focused training and educational program across the educational spectrum include:

- National activities allowing better coordination among all parties to implement a national education and training infrastructure for wind technologies;
- Activities targeting primary and secondary students to expand engagement in energy-related issues and the STEM (science, technology, engineering, and math) fields;
- Efforts to expand community college and vocational training programs and educational standards for all wind technology areas; and
- Consistent and prolonged endeavors to support wind-focused academic institutions and activities at the university and postgraduate levels to ensure a healthy population of wind power professionals with a wide range of expertise, including the sciences, engineering, law, and business.

Numerous efforts are underway to support and expand wind industry workforce development options and to better understand the wind industry's workforce development needs. Various industry groups and educational organizations have already implemented workforce development programs. Activities are also supported by DOE's Wind and Water Power Technologies Office, the American Wind Energy Association, the U.S. Department of Labor, and the National Science Foundation. Many of these efforts are uncoordinated, however, with few direct ties to defined levels of expertise. One of the first actions to support the *Wind Vision Study Scenario* is to improve understanding and coordination of the workforce and educational needs for the wind sector, particularly among academia and industry stakeholders.

The active engagement of students at the primary and secondary levels not only introduces more people to the impacts and benefits of wind power, but also "primes the pump" of the wind power workforce at all levels. Opportunities in STEM topics, including energy and wind technologies, should be made available to students at the K–12 level so they will have the skills and interest to possibly enter the renewable energy workforce. Jobs resulting from these areas of study may be technical, but opportunities exist in policy, regulation, communications, finance, and other support activities.

Because the majority of wind power jobs are supported by community college and vocational level education, a common core of industry-wide job accreditation standards and implementation programs at these levels and in technical centers is essential. Worker education and safety instruction are also critical. Safety certifications for land-based and offshore wind differ, so targeted education and safety guidance are necessary for both. The expanding wind market will require creation of a framework for wind O&M technicians, with particular focus relating to offshore wind development. In addition, clear pathways should be made available for short-service construction workers (land-based and offshore) and vessel operators (offshore) to obtain training and certification related to wind. The industry requires broader, facilitated collaboration to ensure universal understanding of the required skills and defined achievement levels. This understanding will help improve quality of the overall workforce as well as enhance worker flexibility and development. For distributed wind, continued expansion of the market requires more trained site assessors, installers, and maintenance providers.

Many of the skills required for the long-term success of the wind industry, from engineering to business, require individuals with advanced degrees. This need was discussed in a 2013 study by Leventhal and Tegen [37]. Specific actions required include the development of a sustainable university consortium to support R&D efforts; technical training and student collaboration; implementation of an international academic network; creation of sustainable wind-focused university programs; and expansion of opportunities for student, industry, and university collaborations such as internships, research fellowships, and joint research projects.

4.9 Policy Analysis

Wind power offers social benefits and plays a valuable role in the nation's diverse portfolio of electricity generation technologies, but also has potential impacts and faces competition from other electricity generation technologies. National, state, and local policy and regulatory decisions made today and into the future play a significant role in determining the growth of wind power.

Achieving wind power deployment to fulfill national energy, societal, and environmental goals—while minimizing the cost of meeting those goals—is likely to require practical and efficient policy mechanisms that support (directly or indirectly) all three wind power markets: land-based, offshore, and distributed. Objective and comprehensive evaluation of different policy mechanisms is therefore needed, as are comparative assessments of the costs, benefits and impacts of various energy technologies. Regular assessment of progress to enable ongoing prioritization of roadmap actions is also essential.

This section discusses three key areas in which the wind stakeholder community can collaborate with others to maintain the analysis capability necessary to inform policy decision makers:

- **Comprehensively evaluate the costs, benefits, and impacts of energy technologies;**
- **Refine and apply policy analysis methods; and**
- **Track technology advancement and deployment progress and update the roadmap.**

Action 9.1: Refine and Apply Energy Technology Cost and Benefit Evaluation Methods

Thorough evaluation of the costs, benefits, and impacts for all electricity generation alternatives is needed to help guide policy decisions and approaches to achieve societal goals. Historically, comparative evaluations have been based primarily on performance characteristics and direct costs. Various external factors that are not always reflected in direct costs—such as health, water, climate, economic development, and diversity impacts, as well as local impacts on ecosystems and humans—have often not been explored in detail.

ACTION 9.1: Refine and Apply Energy Technology Cost and Benefit Evaluation Methods	
Refine and apply methodologies to comprehensively evaluate and compare the costs, benefits, risks, uncertainties, and other impacts of energy technologies.	
DELIVERABLE	**IMPACT**
A set of recognized and approved methodologies to objectively evaluate the costs, benefits, and impacts of energy technologies, in concert with regular application of these tools.	Increased decision maker access to comprehensive, comparative energy information.

Key Themes: Increase Economic Value for the Nation
Markets Addressed: Land, Offshore, Distributed

Chapter 3 quantifies many of these cost and benefit impacts for wind using best available methods. Additional comprehensive methods are needed for quantifying the full spectrum of costs, benefits, and impacts for all generation options, as well as relative risks and their impacts. These methods would ideally consider various attributes and impacts, and would do so at different geographic and time scales. In some cases, methods to quantify specific impacts do not exist; in other cases, methods exist but are not comprehensively or consistently applied. Further challenge comes in comparing seemingly incommensurable impacts (e.g., comparing bird deaths from wind turbines to air pollution from fossil energy plants), or in determining the specific costs, benefits, and impacts that are appropriate to consider in any particular decision (e.g., carbon effects might be appropriate to consider in national policymaking, but may not be relevant in a local siting decision for a specific wind project). As such, there is a need and opportunity for stakeholder engagement and collaboration not only in methods development, but also in supporting the proper application of those methods by decision makers at the national, state, and local levels.

To become commonly used and accepted, tools will need to be unbiased, with input and buy-in from a wide array of stakeholders. Federal agencies, national laboratories, and academic institutions may be particularly well-positioned to meet this need for comprehensive and unbiased analysis.

Action 9.2: Refine and Apply Policy Analysis Methods

Ongoing reviews of energy and environmental policies are required to evaluate existing policies and enable course corrections as needed, as well as to assess the potential impacts of proposed policies to determine whether they will achieve desired outcomes. A key need and opportunity is to better understand the relative advantages and disadvantages of policy mechanisms that might be used to support renewable energy such as wind, as well as to achieve broader societal goals. The wind community needs to stay abreast of existing and proposed policy options at both the federal and state levels.

Several policies have been used to directly encourage wind power deployment: tax incentives at the federal level, renewable portfolio standards at the state level, and targeted incentive programs for distributed and offshore wind applications. Other types of policy mechanisms under consideration or already in limited use include federal renewable or clean energy portfolio standards, programs to reduce the cost of wind project financing, and policy mechanisms to control the release of greenhouse gas emissions.

Some of the data, models, and tools needed to provide objective energy and environmental policy analysis are already available, but further refinement is an ongoing need, particularly as new policy mechanisms are proposed. One specific need and stakeholder opportunity is to ensure that modeling tools used to evaluate policy options at the federal and state level are able to capture the unique geographic and operational characteristics of wind technology, as well as evolving technology advancements. This need exists for land-based, offshore, and distributed wind applications.

This need for more advanced tools extends beyond wind power and includes improved representation of individual energy technologies as well as electric system planning and operations. To date, energy policy has often been targeted at specific sectors, such as direct incentive support for renewable energy applications. There is broad recognition in the literature, however, that cost-effective achievement of certain societal goals—such as climate change mitigation and reduction in criteria air pollution emissions—calls for broader application of policies focused on external factors that are not always reflected in direct costs. Such policies might include economy-wide pricing of carbon emissions and environmental regulations that comprehensively limit criteria air pollution.

ACTION 9.2: Refine and Apply Policy Analysis Methods	
Refine and apply policy analysis methodologies to understand federal and state policy decisions affecting the electric sector portfolio.	
DELIVERABLE	**IMPACT**
A set of recognized and approved methodologies to objectively evaluate the economic, environmental, societal, and wind-industry impacts of existing and possible future energy policies, in concert with regular application of these tools.	Increased decision maker access to comprehensive evaluations of energy policy options to achieve wind power deployment in fulfillment of national energy, societal, and environmental goals, while minimizing the cost of meeting those goals.

Key Themes: Increase Economic Value for the Nation
Markets Addressed: Land, Offshore, Distributed

Wind stakeholders can collaborate with others, such as those in the broader energy and environmental sectors, to conduct objective analyses that explore the implications of energy policy development on society and the wind industry. A diverse group of entities will continue to create and apply policy analysis tools, both on a commercial basis and to serve specific stakeholder interests. Federal agencies, national laboratories, and academic institutions may be especially well positioned to meet the need for comprehensive and unbiased tools development and policy analysis.

Action 9.3: Maintain the Roadmap as a Vibrant, Active Process for Achieving the Wind Vision Study Scenario

The *Wind Vision* roadmap is intended to be a living document, continually updated to inform stakeholder engagement as technology evolves, deployment expands, and new challenges arise. Roadmap updates will be used as a means to track progress toward the *Wind Vision Study Scenario*. Stakeholders may revisit and revise the roadmap periodically so that it reflects changing circumstances while driving forward momentum.

ACTION 9.3: Maintain the Roadmap as a Vibrant, Active Process for Achieving the *Wind Vision Study Scenario*	
Track wind technology advancement and deployment progress, prioritize R&D activities, and regularly update the wind roadmap.	
DELIVERABLE	**IMPACT**
Periodically produced publicly available reports tracking technology advancement and deployment progress, as well as updated wind roadmaps.	Systematic evaluation of progress towards increased domestic deployment of wind power and identification of any new challenges to be addressed.

Key Themes: Reduce Wind Costs; Expand Developable Areas; Increase Economic Value for the Nation
Markets Addressed: Land, Offshore, Distributed

An abundance of information can be learned from existing wind installations over time, including performance trends, cost trends, O&M experience, technology developments, and electric system integration experience. Accurate tracking and reporting of this information for all three wind markets provides a valuable record of progress in wind technology and applications, as well as early indication of any issues that require attention. This record can inform deliberations and analysis of deployment, policies, and R&D priorities, as well as provide ongoing perspective on the status of wind deployment in the United States relative to the roadmap. As such, stakeholder effort in assembling a thorough and accurate record of U.S. experience with wind power—in all of its applications—is essential. The wind and electric power sectors will play a major role in providing the relevant data, though third-party entities may be best positioned to aggregate, organize, and publish the information while protecting confidentiality.

A range of options for improving cost effectiveness of wind technology and facilitating the technology's use and acceptance are under consideration in both the public and private sectors of the wind community. Stakeholders can support ongoing refinements to the methods used to evaluate and quantify the relative merits of these options, so that priority can be given to those with the greatest expected benefits for complete wind systems. Wind technology advancement opportunities need to be evaluated and tracked in the context of the entire wind power system (or even the entire electric power system) in order to systematically improve the technology's cost effectiveness. Publicly available reports are needed that explain R&D evaluation and prioritization methods as well as the potential influence of successful R&D efforts on the cost of wind technology. R&D priorities then need to be revisited periodically to account for progress made and changing conditions.

Wind industry involvement is required to produce the relevant data to track wind deployment in the United States and provide critical insight on R&D priorities and roadmap revisions. Stakeholders may consider engaging DOE, in conjunction with its national laboratories, as an unbiased third-party to track progress, evaluate technology advancement programs, and update the roadmap.

Chapter 4 References

[1] *Complex Flow Workshop Report.* DOE/GO-102012-3653. Washington, DC: U.S. Department of Energy, May 2012. Prepared following Complex Flow Workshop, Jan. 17–18, 2012, Boulder, CO. Accessed Dec. 20, 2014: *http://energy.gov/eere/wind/downloads/complex-flow-workshop-report.*

[2] Schreck, S.J.; Lundquist, J.K.; Shaw, W.J. *U.S. Department of Energy Workshop Report: Research Needs for Wind Resource Characterization.* NREL/TP-500-43521. Golden, CO: National Renewable Energy Laboratory, 2008. Accessed Jan. 21, 2015: *http://go.usa.gov/JGf9.*

[3] Churchfield, M.J.; Lee, S.; Moriarty, P.J.; Martinez, L.A.; Leonardi, S.; Vijayakumar, G.; Brasseur, J.G. "Large-Eddy Simulation of Wind-Plant Aerodynamics." Preprint prepared for 50th AIAA Aerospace Sciences Meeting, Jan. 9–12, 2012. NREL/CP-5000-53554. Golden, CO: National Renewable Energy Laboratory, 2012; 19 pp. Accessed Jan. 27, 2015: *http://bit.ly/1yLLzC6.*

[4] "NREL High Performance Computing: Mission and Overview." Golden, CO: National Renewable Energy Laboratory, 2014. Accessed Dec. 19, 2014: *https://hpc.nrel.gov/about.*

[5] Musial, W.; Ram, B. *Large-Scale Offshore Wind Power in the United States: Assessment of Opportunities and Barriers.* NREL/TP-500-40745. Golden, CO: National Renewable Energy Laboratory, 2010. Accessed Jan. 27, 2015: *http://www.nrel.gov/wind/publications.html.*

[6] Dedrick, J.; Kraemer, K.L. *Value Capture in the Global Wind Energy Industry.* Irvine, CA: University of California, Irvine, Personal Computing Industry Center, June 2011. Accessed Dec. 19, 2014: *http://escholarship.org/uc/item/7d12t1zc.*

[7] *U.S. Wind Energy Manufacturing and Supply Chain: A Competitiveness Analysis.* DE-EE-0006102. Work performed by Global Wind Network, Cleveland, OH. Washington, DC: U.S. Department of Energy, June 2014. Accessed Oct. 10, 2014: *http://energy.gov/eere/downloads/us-wind-energy-manufacturing-supply-chain-competitiveness-analysis.*

[8] *Global Manufacturing: Foreign Government Programs Differ in Some Key Respects from Those in the United States.* Report to Chairman, Senate Committee on Commerce, Science, and Transportation. GAO-13-365. Washington, DC: U.S. Government Accountability Office, 2013. Accessed Jan. 21, 2015: *http://www.gao.gov/products/GAO-13-365.*

[9] Berger, S. *Making in America: From Innovation to Market.* Cambridge, MA: MIT Press, 2013.

[10] Orrell A. and H.E. Rhoads-Weaver. 2013 Distributed Wind Market Report. PNNL-23484. Richland, WA: Pacific Northwest National Laboratory, 2014. Accessed Jan. 29, 2015: *http://www.energy.gov/eere/wind/downloads/2013-distributed-wind-market-report.*

[11] *World Market Update 2012: International Wind Energy Development Forecast 2013–2017.* Copenhagen: Navigant Research, 2013. Accessed Jan. 27, 2014: *http://www.navigantresearch.com/newsroom/navigant-research-releases-new-btm-wind-report-world-market-update-2012.*

[12] Cotrell, J.; Stehly, J.; Johnson, J.; Roberts, J.O.; Parker, Z.; Scott, G.; Heimiller, D. *Analysis of Transportation and Logistics Challenges Affecting the Deployment of Larger Wind Turbines.* NREL/TP-5000-61063. Golden, CO: National Renewable Energy Laboratory, 2014. Accessed Dec. 19, 2014: *http://energy.gov/eere/wind/downloads/analysis-transportation-and-logistics-challenges-affecting-deployment-larger.*

[13] Elkinton, C.; Blatiak, A.; Ameen, H. *Assessment of Ports for Offshore Wind Development in the United States.* Document 700694-USPO-R-03. Work performed by GL Garrad Hassan America, San Diego, CA. Washington, DC: U.S. Department of Energy, March 2014. Accessed Dec. 19, 2014: *http://energy.gov/eere/wind/downloads/wind-offshore-port-readiness.*

[14] *U.S. Offshore Wind Manufacturing and Supply Chain Development.* Work performed by Navigant Consulting, Burlington, MA. Washington, DC: U.S. Department of Energy, 2013. Accessed Dec. 19, 2014: *http://energy.gov/eere/wind/downloads/us-offshore-wind-manufacturing-and-supply-chain-development.*

[15] *Assessment of Vessel Requirements for the U.S. Offshore Wind Sector.* Work performed by Douglas-Westwood, Kent, UK, for U.S. Department of Energy (DOE). Washington, DC: U.S. DOE, 2013. Subtopic 5.2 of U.S. Offshore Wind: Removing Market Barriers grant opportunity under Funding Opportunity Announcement 414, 2013. Accessed Dec. 19, 2014: *http://energy.gov/eere/wind/downloads/assessment-vessel-requirements-us-offshore-wind-sector*.

[16] *Offshore Wind Industrial Strategy: Business and Government Action.* London: HM Government, August 2013. Accessed Dec. 19, 2014: *www.gov.uk.*

[17] *Offshore Wind Cost Reduction Pathways Study.* London: The Crown Estate, May 2012. Accessed Dec. 19, 2014: *http://www.thecrownestate.co.uk/energy-and-infrastructure/offshore-wind-energy/working-with-us/strategic-workstreams/cost-reduction-study/.*

[18] *Offshore Wind Project Timelines 2013.* London: Renewable UK and Crown Estate, 2013. Accessed Dec. 20, 2014: *http://www.renewableuk.com/download.cfm?docid=63B303B4-425D-4CD3-B032A0A4F109E42C.*

[19] *Onshore Wind Turbines Life Extension.* Product 1024004. Palo Alto, CA: Electric Power Research Institute, October 2012. Accessed Oct. 10. 2014: *http://www.epri.com/abstracts/Pages/ProductAbstract.aspx?ProductId=000000000001024004.*

[20] *AWEA Operation and Maintenance Recommended Practices.* Washington, D.C.: American Wind Energy Association, 2013. Accessed Dec. 19, 2014: http://www.awea.org/oandm.

[21] Kirby, B.; Milligan, M. "The Impact of Balancing Area Size, Obligation Sharing, and Energy Markets on Mitigating Ramping Requirements in Systems with Wind Energy." *Wind Engineering* (32:4), 2008: pp. 379–398. Accessed Dec. 19, 2014: *http://dx.doi.org/10.1260/0309-524X.32.4.379.*

[22] Ela, E.; Gevorgian, V.; Tuohy, A.; Kirby, B.; Milligan, M.; O'Malley, M. "Market Designs for the Primary Frequency Response Ancillary Service—Part I: Motivation and Design." *IEEE Transactions on Power Systems* (29:1), 2014; pp. 421–431. Accessed Oct. 10, 2014: *http://dx.doi.org/10.1109/TPWRS.2013.2264942.*

[23] Ela, E.; Gevorgian, V.; Tuohy, A.; Kirby, B.; Milligan, M.; O'Malley, M. "Market Designs for the Primary Frequency Response Ancillary Service—Part II: Case Studies." *IEEE Transactions on Power Systems* (29:1), 2014; pp. 432-440. Accessed Oct. 10, 2014: *http://dx.doi.org/10.1109/TPWRS.2013.2264951.*

[24] *Competitive Renewable Energy Zone Program Oversight, CREZ Progress Report No. 13 (Oct Update).* Austin, TX: Public Utility Commission of Texas, 2013. Accessed Jan. 27, 2015: *www.texascrezprojects.com/quarterly_reports.aspx.*

[25] Wiser, R.; Bolinger, M. *Annual Report on U.S. Wind Power Installation, Cost, and Performance Trends: 2007.* DOE/GO-102008-2590. Washington, D.C.: U.S. Department of Energy, 2008. Accessed Jan. 27, 2015: *http://www.osti.gov/scitech/biblio/929587.*

[26] Wiser, R.; Bolinger, M. *2008 Wind Technologies Market Report.* DOE/GO-102009-2868. Washington, D.C.: U.S. Department of Energy, 2009. Accessed Jan. 26, 2015: *http://www1.eere.energy.gov/library/default.aspx?Page=9.*

[27] Wiser, R.; Bolinger, M. *2009 Wind Technologies Market Report.* DOE/GO-102010-3107. Washington, D.C.: U.S. Department of Energy, 2010. Accessed Jan. 26, 2015: *http://www1.eere.energy.gov/library/default.aspx?Page=9.*

[28] Wiser, R.; Bolinger, M. *2010 Wind Technologies Market Report.* DOE/GO-102011-3322. Washington, D.C.: U.S. Department of Energy, 2011. Accessed Jan. 26, 2015: *http://www1.eere.energy.gov/library/default.aspx?Page=9.*

[29] Wiser, R.; Bolinger, M. *2011 Wind Technologies Market Report.* DOE/GO-102012-3472. Washington, D.C.: U.S. Department of Energy, 2012. Accessed Jan. 26, 2015: *http://www1.eere.energy.gov/library/default.aspx?Page=9.*

[30] Wiser, R.; Bolinger, M. *2012 Wind Technologies Market Report.* DOE/GO-102013-3948. Washington, D.C.: U.S. Department of Energy, 2013. Accessed Jan. 26, 2015: *http://www1.eere.energy.gov/library/default.aspx?Page=9.*

[31] Wiser, R.; Bolinger, M. *2013 Wind Technologies Market Report.* DOE/GO-102014-445. Washington, D.C.: U.S. Department of Energy, 2014. Accessed Jan. 26, 2015: *http://www1.eere.energy.gov/library/default.aspx?Page=9.*

[32]   Chang, J.W.; Pfeifenberger, J.P.; Hagerty, J.M. *The Benefits of Electric Transmission: Identifying and Analyzing the Value of Investments.* Work performed by The Brattle Group, Boston, MA. Washington, DC: Working Group for Investment in Reliable and Economic Electric Systems, July 2013. Accessed Dec. 19, 2014: *http://www.brattle.com/news-and-knowledge/ news/20.*

[33]   Milligan, M.; King, J.; Kirby, B.; Beuning, S. "Impact of Alternative Dispatch Intervals on Operating Reserve Requirements for Variable Generation." Prepared for 10th International Workshop on Large-Scale Integration of Wind (and Solar) Power into Power Systems as well as on Transmission Networks for Offshore Wind Power Plants, Oct. 25–26, 2011, Aarhus, Denmark. NREL/CP-5500-52506. Golden, CO: National Renewable Energy Laboratory. Accessed Jan. 8, 2015: *http://www.windintegrationworkshop.org/ berlin2014/old_proceedings.php.*

[34]   "Standard for Interconnecting Distributed Resources with Electric Power Systems." IEEE 1547 Working Group, 2014. Accessed Dec. 20, 2014: *http:// grouper.ieee.org/groups/scc21/1547/1547_index.html.*

[35]   *AWEA U.S. Wind Industry Annual Market Report.* American Wind Energy Association. Washington, DC: AWEA, 2014a. Accessed Dec. 13, 2014: *http://www. awea.org/AMR2013.*

[36]   *Annual Energy Outlook 2014.* DOE/EIA-0383 (2014). Washington, DC: U.S. Department of Energy, Energy Information Administration, 2014. Accessed Dec. 14, 2014: *http://www.eia.gov/forecasts/aeo/.*

[37]   Leventhal, M.; Tegen, S. *A National Skills Assessment of the U.S. Wind Industry in 2012.* NREL/ TP-7A30-57512. Golden, CO: National Renewable Energy Laboratory, 2013. Accessed Dec. 19, 2014: *http://www.nrel.gov/docs/fy13osti/57512.pdf.*